BASIC CONCEPTS IN CHEMISTRY

Organic Synthetic Methods

JAMES R. HANSON

University of Sussex

Cover images © Murray Robertson/visual elements 1998–99, taken from the
109 Visual Elements Periodic Table

For ordering and customer service, call 1-800-CALL-WILEY.

Library of Congress Cataloging-in-Publication Data:
Library of Congress Cataloging-in-Publication Data is available.
ISBN: 0-471-54910-X

Typeset in Great Britain by Wyvern 21, Bristol
Printed and bound by Polestar Wheatons Ltd, Exeter

10 9 8 7 6 5 4 3 2 1

Preface

A successful synthesis of an organic compound requires a sound grasp of functional group chemistry, reaction mechanisms and stereochemistry in order for the student to be able to understand the methods of making bonds. It then requires a good grasp of their structural consequences in order to analyse a target molecule in terms of the bonds that can be disconnected to reveal suitable synthetic steps. A retrosynthetic analysis of the target molecule must involve plausible intermediates and suggest readily available starting materials. This analysis rests both on a familiarity with common building blocks and synthetic strategies.

This book is aimed at the second year undergraduate who has already completed a course of functional group chemistry. The first chapters of the book describe the ways of making carbon–carbon bonds. These are followed by a chapter on making carbon–nitrogen bonds. Then follows functional group interconversions involving oxidation, reduction and halogenation, which play important roles in synthesis. The control over the regioselectivity of reactions exerted by protecting groups is then described. Finally, the strategies that have led to some syntheses are outlined.

A balance has to be drawn between the use of the systematic and the trivial names for compounds. Whilst systematic names have been given for some compounds, many reagents are known by their trivial names or by abbreviations based on them. Hence the trivial, non-systematic names are used in these cases.

I am indebted to Martyn Berry, Professor Alwyn Davies FRS and to Professor Sir John Cornforth AC FRS for their substantial help in the preparation of the manuscript.

James R. Hanson
University of Sussex

BASIC CONCEPTS IN CHEMISTRY

This series of books consists of short, single-topic or modular texts, concentrating on the fundamental areas of chemistry taught in undergraduate science courses. Each book provides a concise account of the basic principles underlying a given subject, embodying an independent-learning philosophy and including worked examples. The one topic, one book approach ensures that the series is adaptable to chemistry courses across a variety of institutions.

TITLES IN THE SERIES

Stereochemistry *D G Morris*
Reactions and Characterization of Solids *S E Dann*
Main Group Chemistry *W Henderson*
d- and f-Block Chemistry *C J Jones*
Structure and Bonding *J Barrett*
Functional Group Chemistry *J R Hanson*
Organotransition Metal Chemistry *A F Hill*
Heterocyclic Chemistry *M Sainsbury*
Atomic Structure and Periodicity *J Barrett*
Thermodynamics and Statistical Mechanics *J M Seddon & J D Gale*
Basic Atomic and Molecular Spectroscopy *J M Hollas*
Organic Synthetic Methods *J R Hanson*
Aromatic Chemistry *J D Hepworth, D R Waring and M J Waring*
Quantum Mechanics for Chemists *D O Hayward*
Peptides and Proteins *S Doonan*

Further information about this series is available at www.wiley.com/go/wiley-rsc

Contents

1 Introduction **1**

1.1 Synthetic Strategies 1
1.2 Steric Factors 3
1.3 Criteria for Evaluating Synthetic Routes 3
1.4 Some Synthetic Terms 6
1.5 Learning to Apply Synthetic Methods 6

2 Organometallic and Ylide Methods of Carbon–Carbon Bond Formation **9**

2.1 Introduction 9
2.2 Reactions of Organometallic Compounds 11
2.3 Acetylides and Nitriles 20
2.4 Ylide Reactions 22
2.5 Silicon and Boron in C–C Bond Formation 26

3 Carbonyl Activation and Enolate Chemistry in Carbon–Carbon Bond Formation **32**

3.1 Introduction 32
3.2 Alkylation Reactions 36
3.3 Enolate Anions in Carbonyl Addition Reactions 39
3.4 The Stereochemistry of Condensation Reactions 48

4 Carbocations in Synthesis 54

4.1 Introduction 54
4.2 Alkyl Carbocations: the Friedel–Crafts Alkylation 55
4.3 Carbocations Derived from Aldehydes and Ketones 56
4.4 Acylium Carbocations: the Friedel–Crafts Acylation 57
4.5 Acid-catalysed Rearrangement Reactions 58

5 Free Radical and Pericyclic Reactions in the Formation of Carbon–Carbon Bonds 64

5.1 Carbon Radical Reactions 64
5.2 Radical Addition Reactions 68
5.3 Carbenes 69
5.4 Alkene Metathesis 70
5.5 The Diels–Alder Reaction 71
5.6 The Ene Reaction 73
5.7 The Cope and Claisen Rearrangements 73

6 Methods of Making Carbon–Nitrogen Bonds 78

6.1 Introduction 78
6.2 Electrophilic Methods of Making C–N Bonds 79
6.3 Nucleophilic Methods of Making C–N Bonds 80
6.4 Rearrangement Methods 84
6.5 The Synthesis of Amino Acids 85
6.6 The Synthesis of Heterocyclic Compounds 87

7 Functional Group Transformations 96

7.1 Oxidation 96
7.2 Reduction 103
7.3 Halogenation 111

8 Protecting Groups 126

8.1 Protection of Functional Groups 126
8.2 Peptide Synthesis 133
8.3 Combinatorial Synthesis 136

9 Some Examples of Total Syntheses 142

9.1 Introduction 142
9.2 β-Eudesmol 143
9.3 Griseofulvin 144
9.4 Thiamine (Vitamin B_1) 147
9.5 Prostaglandins 150

Further Reading 153

Answers to Problems 157

Subject Index 173

1
Introduction

Aims

The aim of this chapter is to outline the various strategies for designing a synthesis and evaluating the proposed synthetic schemes. By the end of this chapter you should appreciate:

- The concept of the retrosynthetic analysis of a target molecule
- The role that the relationship between functional groups in a target molecule plays in designing a synthesis
- The importance of identifying basic building blocks in a target molecule
- The criteria by which a synthetic scheme might be evaluated

1.1 Synthetic Strategies

Organic synthesis represents one of the major activities of organic chemists. The reasons for undertaking a synthesis are varied. The unambiguous synthesis of a compound such as a natural product provides the ultimate confirmation of its structure. Many syntheses are undertaken to explore the physical, chemical or biological properties of a compound. Other syntheses, particularly of labelled compounds, have the object of examining the metabolism of a compound. In this case the synthetic route that is selected may be constrained by the need to introduce the label at a particular site. Finally, a synthesis may be undertaken in order to demonstrate the scope of a new synthetic methodology. In each case, a synthetic route must be devised that fulfils various criteria of yield, efficiency and flexibility.

There are a number of general strategies for the design of a synthesis leading to a specific target molecule. The first of these involves the

retrosynthetic analysis of the target molecule. This must be based on a sound knowledge of functional group chemistry and the structural consequences of reactions. The bonds of the target molecule are systematically examined to identify particular reactions that might be used in their construction. The relationship between the functional groups in a target molecule may reveal particular synthetic strategies. It is sensible to examine each bond attached to a functional group and to ask the question "Is there a good way to make this bond?" The bonds that are formed in these steps are disconnected to reveal the structures of precursor molecules, which are then examined in a similar way. The key bonds that can be disconnected to reveal the synthetic strategy are known as the strategic bonds. Their identification provides the clue to a successful synthesis. This analysis may lead to a series of **disconnections** and to a "family tree" of subsidiary target structures. Sometimes, in order to reveal a disconnection it may be necessary to modify a functional group by oxidation or reduction.

Useful aids to identifying disconnections are to look at the relationship between functional groups, to look for elements of symmetry within a molecule, and to remove appendages which might be introduced at a late stage in a synthesis. A further aid to the retrosynthetic analysis of a target molecule is to identify each of the rings. There are well-established strategies for constructing particular ring systems which may be buried within a molecule. Computer programs have been written which carry out this analysis in a systematic way.

The symbol $\Rightarrow$ is used to indicate a disconnection, *i.e.* the reverse of a synthetic reaction. The term "**synthon**" is sometimes used to describe the sub-structure or the reactive component in a synthetic step.

A simple example might be to design a synthesis of phenacetin. The bonds that might be made are indicated in the structure (Scheme 1.1). This analysis reveals the need not only to consider the bonds that might be made but also to consider the possibility of functional group interconversions. Whereas the direct introduction of an amino group on to an electron-rich aromatic ring is an unlikely process, nitration or nitrosation followed by reduction are feasible ways of achieving this. In the

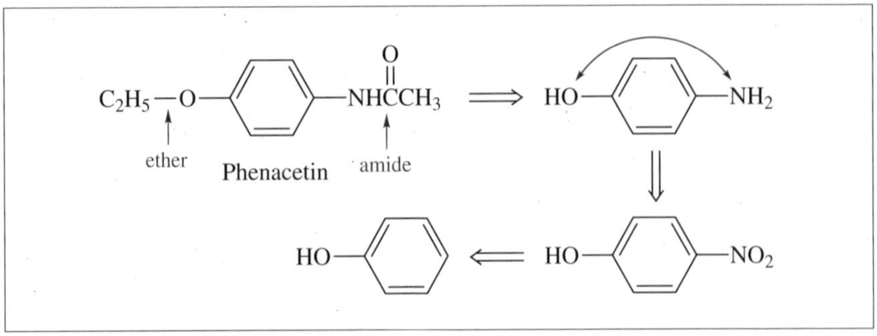

C_2H_5—O—C_6H_4—NHC(=O)CH_3 (Phenacetin; ether, amide) $\Rightarrow$ HO—C_6H_4—NH_2 $\Rightarrow$ HO—C_6H_4—NO_2 $\Rightarrow$ HO—C_6H_5

Scheme 1.1

synthesis there is also a more subtle question. Should the acetyl (ethanoyl) group be introduced before the ethyl group, or *vice versa*? The more nucleophilic nitrogen is acetylated before the phenol is alkylated.

Another method for analysing the target molecule that can reveal a suitable synthetic strategy is to examine the structure for the residue of any of the common building blocks. This is particularly useful when developing a synthesis leading to a labelled molecule, or when the synthetic target is a chiral molecule. The use of an appropriate chiral building block may lead to the eventual formation of one enantiomer of the target molecule. An analysis of the target structure in terms of the building blocks may be important in an industrial context in order to devise a cost-effective synthesis. The synthetic route which is used may be determined by the availability of the starting materials. The dissection of the structure of phenacetin in these terms is given in Scheme 1.2.

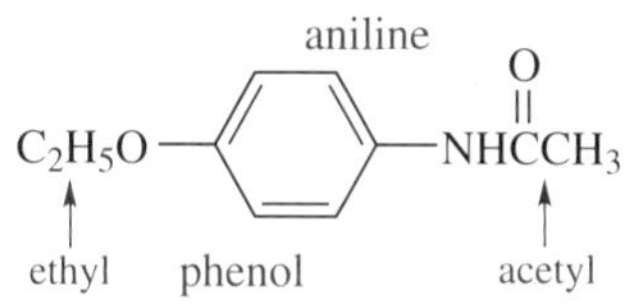

aniline

C_2H_5O—C_6H_4—$NHC(=O)CH_3$

ethyl phenol acetyl

Scheme 1.2

1.2 Steric Factors

The success of a synthesis relies on a combination of favourable electronic and steric factors. In designing a synthesis, careful consideration must be given to obtaining the correct stereoisomers. Hence the retrosynthetic analysis of the target molecule must include an evaluation of the stereochemical features both of the target molecule and of any synthetic reactions implicit in the disconnections. For example, an alkene can show geometrical (*cis*/*trans*) isomerism. Although an alkene may appear at first sight to be a suitable point for a disconnection, the creation of the alkene in a stereospecific manner may present a limitation. An apparently suitable strategic bond may be sterically hindered, and creating a ring junction with the correct stereochemistry may also present a challenge. In the synthesis of a biologically active chiral molecule, obtaining the correct enantiomer is obviously important. Some of these features are shown in the structure of the plant hormone abscisic acid (Scheme 1.3), which has been the target of a number of syntheses.

1.3 Criteria for Evaluating Synthetic Routes

Analyses of the target molecule may suggest a series of possible synthetic routes, and so there have to be criteria for evaluating these. A synthesis

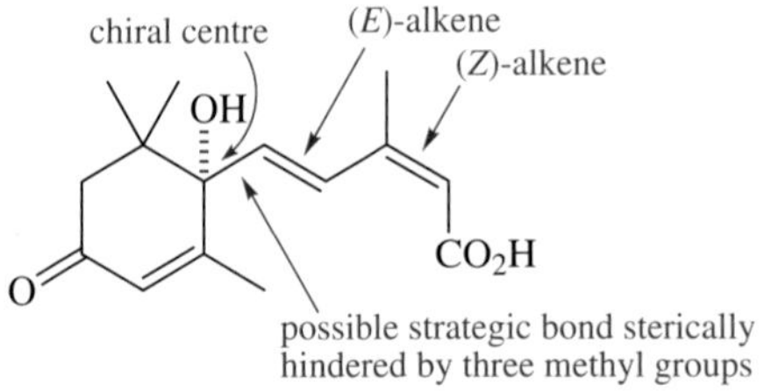

chiral centre (*E*)-alkene (*Z*)-alkene

OH

CO_2H

O

possible strategic bond sterically hindered by three methyl groups

Scheme 1.3

may follow a **convergent** or a **linear** strategy. In the former, two or more fragments of the molecule are assembled separately and are then brought together at a late stage in the synthesis, whereas in the latter the molecule is constructed in a stepwise fashion. A convergent synthesis can have an advantage over the linear synthesis in terms of yield. The effect of an 80% yield at each stage for two syntheses of a target molecule (TM) is shown in Scheme 1.4.

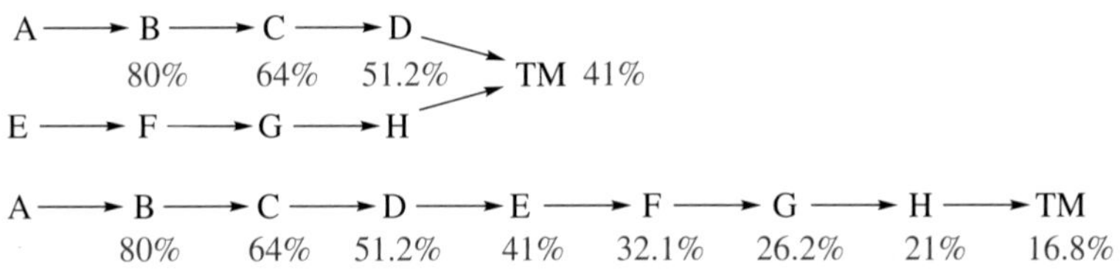

A → B → C → D
80% 64% 51.2%
→ TM 41%
E → F → G → H

A → B → C → D → E → F → G → H → TM
80% 64% 51.2% 41% 32.1% 26.2% 21% 16.8%

Scheme 1.4

A convergent synthesis has greater flexiblity when a series of compounds is being produced for biological testing. One arm can be varied while the other is kept constant. Furthermore, a convergent synthesis may be useful when mutually incompatible or very reactive functional groups are present in a molecule, in that they may be kept apart in different arms of the synthesis. There are also advantages of safety and economy in the preparation of labelled molecules.

Worked Problem 1.1

Q Carry out a retrosynthetic analysis of the anti-inflammatory drug phenylbutazone (**1**).

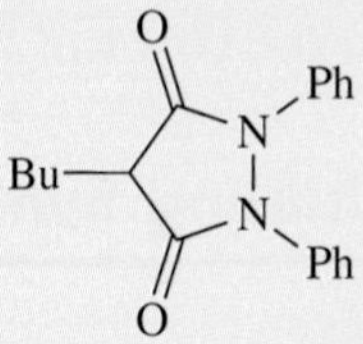

O Ph N N Bu Ph O

1

A The key to this analysis lies in identifying the amide bonds as strategic bonds and secondly in recognizing the symmetry in the target molecule. The first retrosynthetic step suggests a reaction between a diphenylhydrazine and a malonyl chloride (propanedioyl chloride). A functional group interconversion links the malonyl chloride with a malonate ester. The latter may be alkylated with 1-chlorobutane.

$$\Longrightarrow \text{Bu}-\text{CH}(\text{COCl})_2 + \text{PhNH}-\text{NHPh} \Longrightarrow \text{BuCl} + \text{CH}_2(\text{CO}_2\text{Et})_2$$

Worked Problem 1.2

Q Suggest potential building blocks for the painkiller lidocaine (**2**).

$$2,6\text{-Me}_2\text{C}_6\text{H}_3-\text{NHCOCH}_2\text{NEt}_2 \quad \textbf{2}$$

A The structure may be divided into three units: 2,6-dimethylaniline, a $COCH_2$ unit and a diethylamine unit. This synthesis involves the following steps. In this synthesis, note the differing reactivity of the acyl chloride and the alkyl chloride, and secondly the difference in reactivity between the arylamine and the alkylamine which defines the sequence of the reactions.

$$2,6\text{-Me}_2\text{C}_6\text{H}_3-\text{NH}_2 + \text{ClCOCH}_2\text{Cl} \longrightarrow 2,6\text{-Me}_2\text{C}_6\text{H}_3-\text{NHCOCH}_2\text{Cl} + \text{Et}_2\text{NH} \longrightarrow \textbf{2}$$

1.4 Some Synthetic Terms

A number of terms are used to describe reactions that are used in synthesis. A **regiospecific reaction** occurs entirely at one centre, whilst a **regioselective reaction** occurs predominantly at one centre. A **stereospecific reaction** is one which produces a single stereochemical result, whilst in a **stereoselective reaction** one stereoisomer predominates. **Enantiospecific reactions** produce just one enantiomer, whilst **enantioselective reactions** favour the formation of one enantiomer.

Protecting groups are often introduced to mask a reactive centre in order to allow a transformation to be carried out elsewhere in a molecule. **Activating groups** may be introduced to enhance the reactivity of a particular centre, and thus favour the formation of a particular product.

The **total synthesis** of a natural product refers to its synthesis from simple commercially available starting materials, whereas its **partial synthesis** refers to a synthesis from another, often more readily available, natural product.

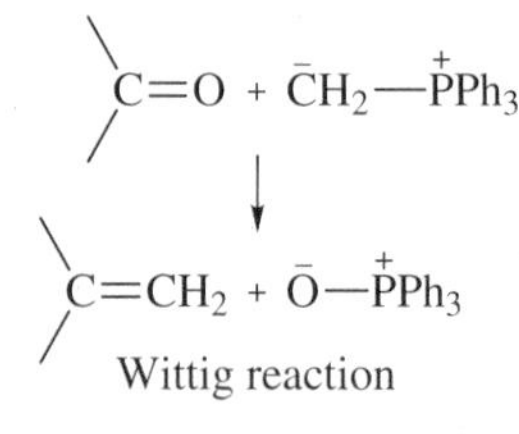

Wittig reaction

Diels–Alder reaction

aldol condensation

1.5 Learning to Apply Synthetic Methods

1. Each of the major synthetic reactions generate specific structural features which act as **markers** for the use of the reaction in a synthesis (*e.g.* an alkene from a Wittig reaction, a *cis* ring junction from a Diels–Alder reaction). As you read through this book, learn to recognize these characteristic features in different structures.
2. The **relationship between functional groups** may provide the clue to a particular strategy (*e.g.* a β-hydroxy ketone may imply an aldol condensation). As you learn a synthetic reaction, note the relationship that is involved between: (i) the activating group that generates the reactive species, (ii) the new bond that is formed and (iii) the structural fate of the recipient group which is attacked by the reactive species.
3. The presence of **electron-donating** or **electron-withdrawing** groups may favour particular reactions at particular centres. For example, this is very important in aromatic substitution. As you learn a synthetic sequence, note the influence of the factors such as resonance that may favour regions of electron density or deficiency.
4. In dissecting a target molecule, look for **competing functional groups** and then either mask (protect) them or place them in a separate arm of a convergent synthesis.
5. Look for the residues of standard **building blocks** [*e.g.* a CH_3CO (Ac) group from ethyl acetoacetate (ethyl 3-oxobutanoate)].

6. A helpful way of learning to dissect simple target molecules and applying synthetic methods is to consider the ways in which particular centres might be isotopically labelled for metabolic studies. The common amino acids and biosynthetic intermediates, such as mevalonic acid lactone (4-hydroxy-4-methyltetrahydro-2*H*-pyran-2-one) and oxaloacetic acid (2-oxosuccinic acid), are good examples in this context.

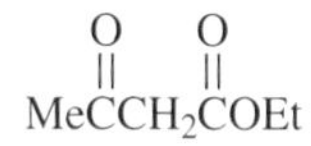

$MeC(=O)CH_2C(=O)OEt$
ethyl acetoacetate

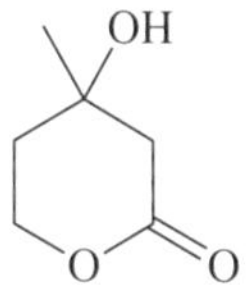

mevalonic acid lactone

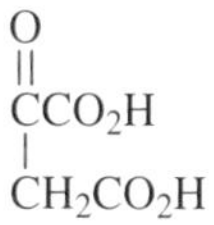

$HO_2CC(=O)CH_2CO_2H$
oxaloacetic acid

In subsequent chapters we will consider methods for making carbon–carbon bonds using carbon nucleophiles, carbon electrophiles, carbon radicals and electrocyclic reactions. We will then consider methods for making carbon–nitrogen bonds, functional group interconversions, and the role of protecting groups in synthesis. Finally, we will examine some examples of total syntheses.

Summary of Key Points

1. The relationship between functional groups in a target molecule may reveal key disconnections in the retrosynthetic analysis.

2. Other disconnections may be revealed by functional group interconversions.

3. Key structural fragments may suggest the use of particular "building blocks" in a synthesis.

4. The identification of particular rings may suggest specific strategies.

5. A convergent synthesis has significant advantages over a linear synthesis.

Problems

1.1. Based on functional group chemistry, suggest possible dissections and building blocks for the target molecules (a)–(f).

(a) $HC{\equiv}C{-}C(Me)(OH){-}Me$

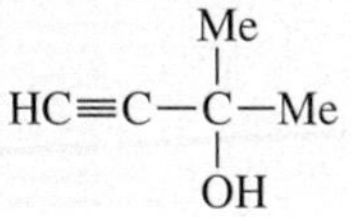

(b) $PhC(=O)NEt_2$

(c) $Me_2CHOC(=O){-}C_6H_4{-}NO_2$ (meta)

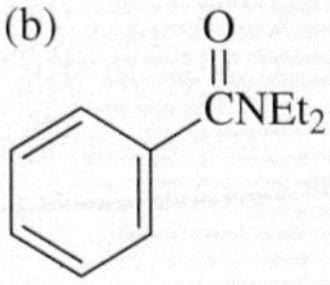

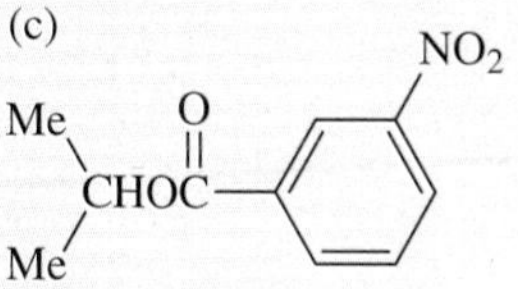

(d) $PhC(=O)CH_2C(=O)OEt$ (e) $PhOCH_2CH(OH)CH_2NMe_2$ (f) $MeC(=O)NH-C_6H_4-SO_2NH_2$

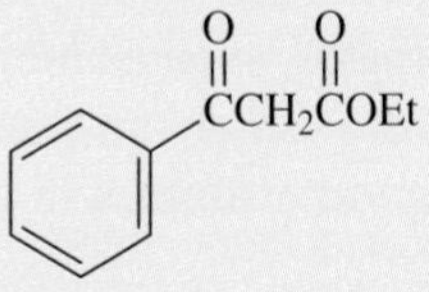

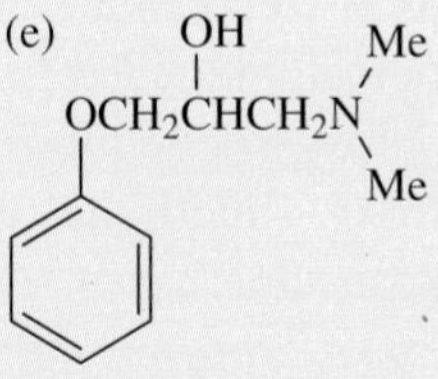

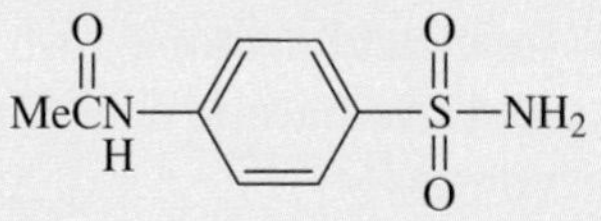

Further Reading

E. J. Corey and X.-M. Cheng, *The Logic of Chemical Synthesis*, Wiley, New York, 1995.

E. J. Corey, *Computer assisted analysis of complex synthetic problems*, in *Q. Rev. Chem. Soc.*, 1971, **25**, 455.

P. A. Wender, S. T. Handy and D. L. Wright, *Towards the ideal synthesis*, in *Chem. Ind. (London)*, 1997, 765.

K. C. Nicolaou, E. J. Sorensen and N. Winssinger, *The art and science of organic natural products synthesis*, in *J. Chem. Educ.*, 1998, **75**, 1225.

K. C. Nicolaou, D. Vourloumis, N. Winssinger and P. S. Baran, *The art and science of total synthesis at the dawn of the 21st century*, in *Angew. Chem. Int. Ed.*, 2000, **39**, 44.

2

Organometallic and Ylide Methods of Carbon–Carbon Bond Formation

Aims

The aim of this chapter is to show how organometallic and ylide methods of generating a carbanion can be used in the formation of carbon–carbon bonds. By the end of this chapter you should understand:

- The major methods of generating organometallic reagents
- The roles of magnesium, lithium, copper, palladium and other metals
- The reactions of ylide reagents and the role of phosphorus and sulfur in these
- The use of silicon in synthesis
- The consequences of the reactions of these species with electron-deficient centres

2.1 Introduction

When carbon is bonded to a more electropositive element such as a metal, the carbon becomes negatively polarized ($C^{\delta-}$–$M^{\delta+}$). In these organometallic compounds the carbon may behave as a nucleophile and then react with an electron-deficient centre. Many organometallic compounds are prepared from alkyl halides (halogenoalkanes) by the insertion of the metal into a carbon–halogen bond. In the starting alkyl halide the carbon is more electropositive than the halogen and the carbon–halogen bond is polarized in the sense $C^{\delta+}$–$hal^{\delta-}$. In the organometallic compound the reverse is the case and there is a reversal in the reactive character of the carbon atom. These compounds are therefore very useful in synthesis. Some organometallic bonds break in a homolytic sense, and the reactions are those of free radicals.

RM (M = Li, MgX, *etc.*)
organometallic compound

$R_3\overset{+}{P}-\overset{-}{C}R_2$
ylide

The value of many organometallic compounds in synthesis depends on the extent to which nucleophilic reactivity arises from the ionic character of the particular metal–carbon bonds. Secondly, the metal can modify the electron-deficient character of the centre at which the carbanion reacts. For example, metals such as magnesium may coordinate to the oxygen of a carbonyl group, increasing the electron deficiency of the carbonyl carbon atom.

The increased "s" character of sp^2 and sp hybridized carbon means that these electrons are held more closely to the nucleus. Consequently, sp^2 and sp carbanions are formed more readily, *i.e.* the acidity ($RH \rightleftharpoons R^- + H^+$) increases in the order CH_3–H (sp^3 C) < CH_2=CH–H (sp^2 C) < HC≡C–H (sp C). Hence vinyl (ethenyl) and alkynyl (ethynyl) organometallic compounds are formed more easily than the corresponding sp^3 compounds.

Typical metals involved in synthetically useful organometallic compounds are lithium, sodium, magnesium, zinc, cadmium, copper and tin. Palladium plays an important catalytic role. Other metals that are less commonly used include aluminium, titanium, chromium, iron, cobalt and nickel. In some instances the organometallic reagent is combined with a second carbanion-stabilizing fragment, for example the trimethylsilylmethyl Grignard reagent Me_3SiCH_2MgCl, in which the silicon stabilizes the Grignard. In the **Reformatski reagent** ($BrZnCH_2CO_2Et$) the carbonyl group stabilizes the carbanion as an enolate [CH_2=C(OZnBr)OEt].

The commonest method of preparation of an organometallic compound involves treatment of an organic halide such as iodomethane with a metal such as magnesium (Scheme 2.1). Although the solvent may at first appear to be unreactive, it can play an important role in stabilizing the organometallic derivative. Thus diethyl ether (ethoxyethane) coordinates to magnesium in the **Grignard reagents**

$Et_2O \rightarrow RMgI \leftarrow OEt_2$ (Et–O–Et coordinated above and below RMgI)

Scheme 2.1

$$MeI + Mg \xrightarrow{Et_2O} MeMgI$$

$$2MeMgI + CdCl_2 \xrightarrow{Et_2O} Me_2Cd + 2MgICl$$

Organometallic compounds may react with other metal salts, leading to an interchange of the metal (Scheme 2.1). This interchange is valuable in varying the reactivity of the carbon in the organometallic derivative.

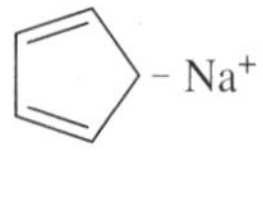

Some C–H bonds are sufficiently acidic to react with metals or metallic bases such as sodamide to form organometallic derivatives. These are exemplified by alkynes and by cyclopentadiene. The cyclopentadienyl anion, with six π-electrons, possesses some aromatic stabilization.

$HC{\equiv}C^-\ Na^+$

$C_5H_5^-\ Na^+$ (cyclopentadienyl sodium)

2.2 Reactions of Organometallic Compounds

2.2.1 Organomagnesium (Grignard) Reagents

Many of the useful synthetic reactions of Grignard reagents involve the addition of the reagent to a carbonyl group. These are shown in Scheme 2.2.

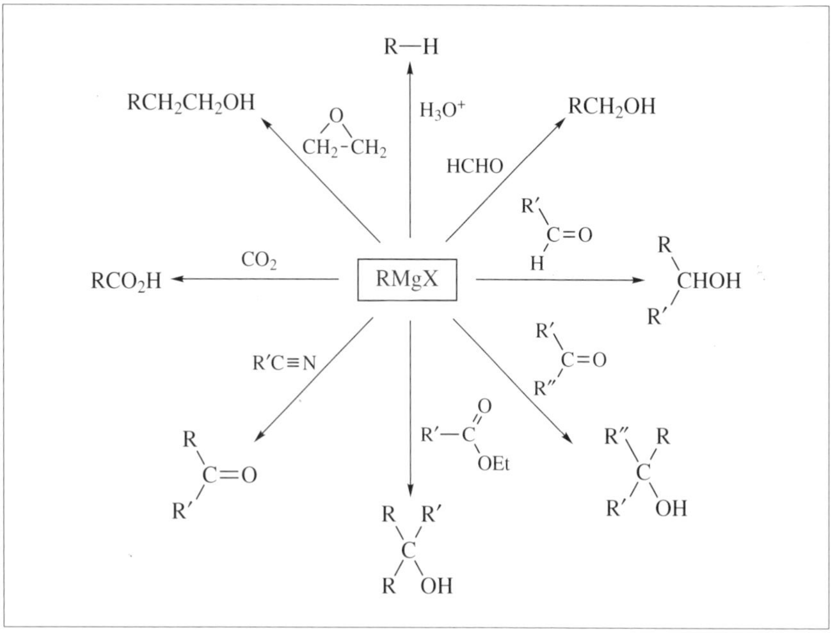

R—H
H_3O^+
RCH_2CH_2OH
CH_2-CH_2 (O)
HCHO
RCH_2OH
RCO_2H
CO_2
RMgX
R′HC=O
RR′CHOH
R′C≡N
RR′C=O
R′R″C=O
R′—C(=O)OEt
RR′R″C—OH
RR′RC—OH

Scheme 2.2

Most of these reactions give an alcohol, and so in carrying out a retrosynthetic analysis of a target molecule the presence of an alcohol can be a **marker** for the use of these reactions. However, the reaction with carbon dioxide stops at the carboxylic acid stage, because the product is the carboxylate (RCO_2^- MgX^+) in which the anion is resistant to further attack by the nucleophile (R^-) derived from the Grignard reagent.

A number of methods have been devised to halt the Grignard addition reactions at intermediate stages to allow the preparation of an aldehyde or a ketone. These reactions involve the formation of a masked carbonyl group such as an amino acetal or an imino anion. The aldehyde or ketone is then only released during the work-up of the reaction (Scheme 2.3).

Grignard reagents react with epoxides to form alcohols in which the new carbon–carbon bond is in the β-position to the hydroxyl group. With an epoxide such as epoxyethane (ethylene oxide), this leads to the extension of a carbon chain by two carbon atoms.

Grignard reagents are important in making other organometallic

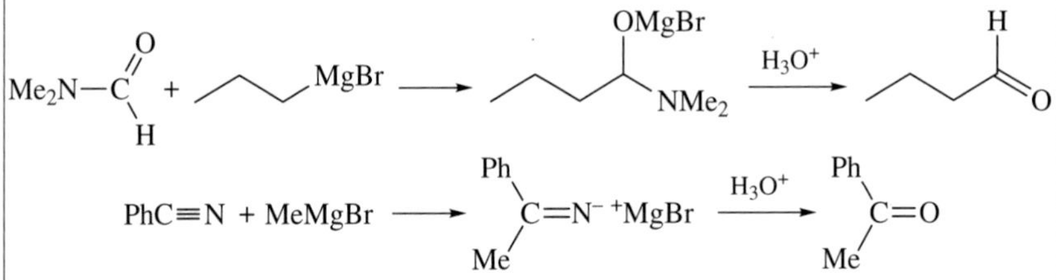

Scheme 2.3

reagents. For example, reaction with silicon tetrachloride ($SiCl_4$) gives rise to di- and trialkylsilyl chlorides and to the fully alkylated silane (R_4Si). These reactions are used to prepare chlorotrimethylsilane (a reagent for introducing the trimethylsilyl protecting group) and for making the NMR standard, tetramethylsilane. Triphenylphosphine is prepared from phosphorus trichloride and phenylmagnesium bromide.

Me_3SiCl
Chlorotrimethylsilane
Me_4Si
Tetramethylsilane
Ph_3P
Triphenylphosphine

When a carbonyl group is sterically hindered, addition of the Grignard reagent may not occur and instead a reduction takes place. 2,4-Dimethylpentan-3-one (diisopropyl ketone) contains the bulky isopropyl groups. Treatment of this ketone with the bulky Grignard reagent propan-2-ylmagnesium bromide gives the alcohol 2,4-dimethylpentan-3-ol, by a reduction (Scheme 2.4). However, the addition reaction succeeds with the more reactive, sterically less-demanding, organolithium compound.

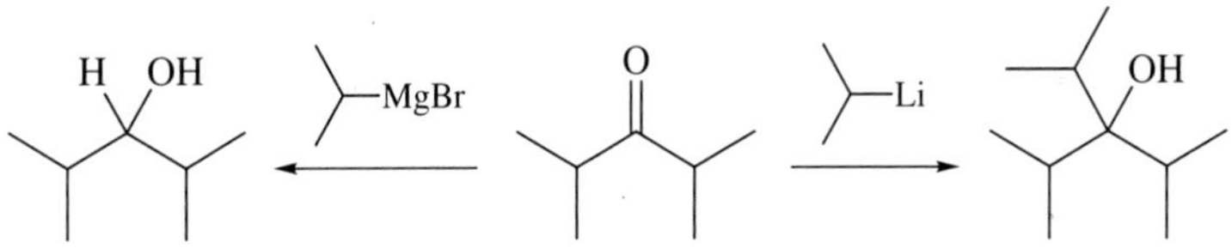

Scheme 2.4

Worked Problem 2.1

Q Propose a synthesis of the labelled alcohol **1**.

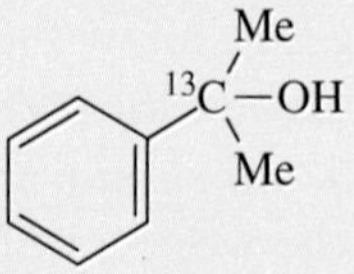

A The presence of a tertiary alcohol in the target molecule, with two identical groups attached to it, suggests the use of a Grignard reaction of an ester in the synthesis. The synthetic scheme is the following:

If a label (*e.g.* deuterium) on the methyl groups was required, a Grignard reaction could be carried out using CD_3I.

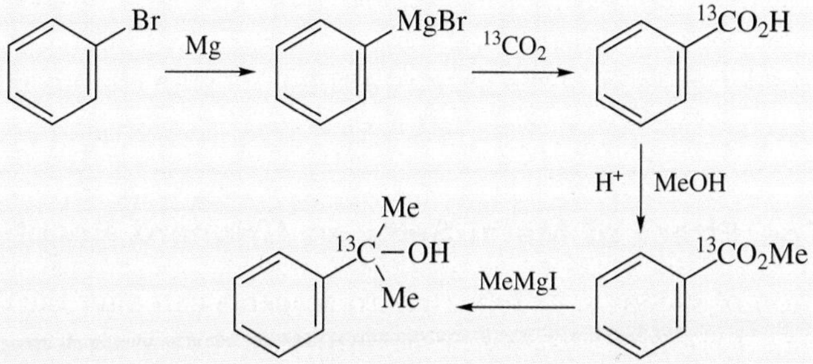

2.2.2 The Stereochemistry of Grignard Reactions

In an open-chain system in which there is free rotation about the carbon–carbon bonds, the relative sizes and stereochemistry of the different substituents on the carbon atom adjacent to the carbonyl group can affect the stereochemistry of addition to the carbonyl group. These groups may be designated S (small), M (medium) and L (large). The preferred conformation may be determined either by the interactions of the carbonyl oxygen or by those of the large group (L) and the alkyl residue (R).

In the situations in which the interactions of the carbonyl oxygen determine the conformation and in which it may be complexed to a Lewis acid, for example a magnesium salt, the carbonyl group takes up a conformation in which the oxygen is *anti* to the large group. The nucleophile then attacks the electron-deficient carbon from the side of the smallest substituent.

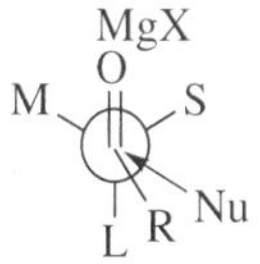

If there is complexing or hydrogen bonding involving an adjacent substituent and the carbonyl oxygen, then this will determine the preferred conformation (Scheme 2.5).

When the nature of the large group L determines the outcome, the preferred conformation may be that in which this group is equidistant from the "R" group and the carbonyl oxygen (**Felkin–Anh model**). The incoming nucleophile attacks the carbonyl group from the less-hindered face, approaching the carbon atom at an angle of about 107° to the C–O bond.

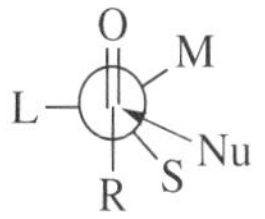

When the carbonyl group is in a ring system, the approach of the reagent will be determined by interactions with the substituents on the ring. Methylmagnesium halides react with cyclic ketones with a very pro-

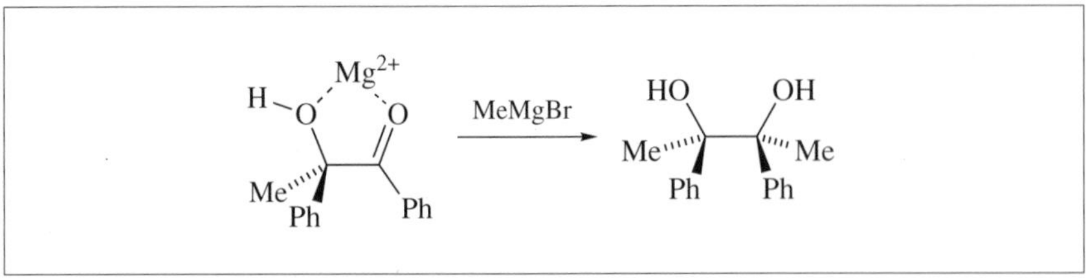

$H-O\cdots Mg^{2+}\cdots O$ (chelate of Me/Ph hydroxy ketone) $\xrightarrow{MeMgBr}$ HO, OH, Me, Me, Ph, Ph diol

Scheme 2.5

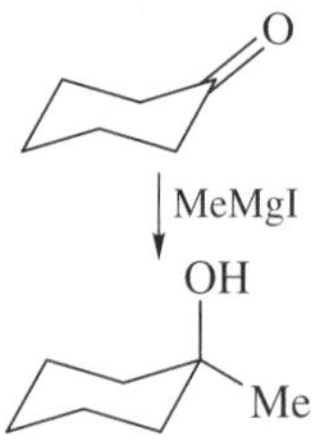

O

MeMgI

OH

Me

nounced tendency to form the tertiary alcohol with an equatorial methyl group.

2.2.3 The Effect of Metal Salts on Grignard Reactions

The addition of metals salts to Grignard reagents can alter the outcome of reactions. Free radicals may be generated in a Grignard reaction by the addition of **cobalt(II) chloride**. An example of this occurs in the reaction between cinnamyl chloride (3-phenylpropenyl chloride) and methylmagnesium chloride. In the absence of the metal salt, the major product is 1-phenylbut-1-ene, which is formed by nucleophilic substitution of the chlorine; this is accompanied by only a small amount of the dimer, 1,6-diphenylhexa-1,5-diene. Addition of cobalt(II) chloride leads to the formation of 1,6-diphenylhexa-1,5-diene as the major product (Scheme 2.6).

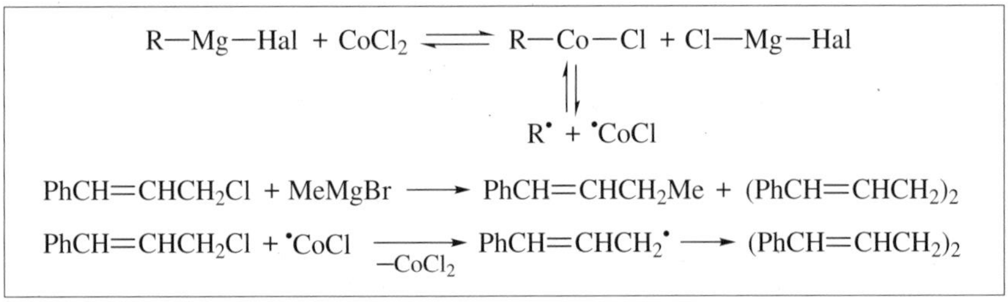

$$R{-}Mg{-}Hal + CoCl_2 \rightleftharpoons R{-}Co{-}Cl + Cl{-}Mg{-}Hal$$

$$R{-}Co{-}Cl \rightleftharpoons R^{\bullet} + {}^{\bullet}CoCl$$

$$PhCH{=}CHCH_2Cl + MeMgBr \longrightarrow PhCH{=}CHCH_2Me + (PhCH{=}CHCH_2)_2$$

$$PhCH{=}CHCH_2Cl + {}^{\bullet}CoCl \xrightarrow[-CoCl_2]{} PhCH{=}CHCH_2^{\bullet} \longrightarrow (PhCH{=}CHCH_2)_2$$

Scheme 2.6

The conversion of methylmagnesium iodide to the **organocadmium** reagent Me_2Cd followed by reaction with an acyl chloride provides a method for the preparation of methyl ketones (Scheme 2.7). Dimethylcadmium does not add to isolated carbonyl groups.

The addition of Grignard reagents to unsaturated ketones is affected by the addition of metal salts. **Copper** salts favour 1,4-addition whilst **cerium** salts favour 1,2-addition. Thus treatment of isophorone (3,5,5-trimethylcyclohex-2-en-1-one) with methylmagnesium iodide on its own gave a 67% yield of the tertiary alcohol, whilst in the presence of copper(I) chloride nearly 83% of the 1,4-addition product was obtained (Scheme 2.7).

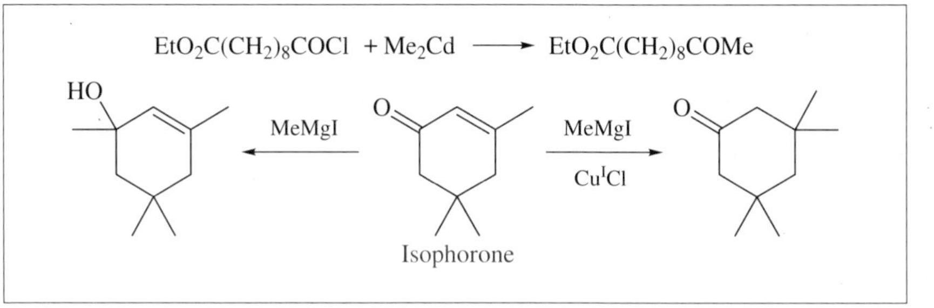

$EtO_2C(CH_2)_8COCl + Me_2Cd \longrightarrow EtO_2C(CH_2)_8COMe$

HO; MeMgI; O; Isophorone; MeMgI, Cu^ICl; O

Scheme 2.7

2.2.4 Organolithium Reagents

Organolithium reagents are prepared by treatment of an alkyl halide with lithium metal or by a metal exchange reaction. Organolithium reagents are more reactive than Grignard reagents and are sterically less demanding, as shown by the reaction of 2,4-dimethylpentan-3-one with isopropyllithium (Scheme 2.4). Another reaction which illustrates their enhanced reactivity is with carbon dioxide. Unlike Grignard reagents, which give a carboxylic acid, the organolithium has sufficient nucleophilicity to add to the lithium carboxylate to give a ketone. The lithium acetal intermediate [$R_2C(OLi)_2$] protects the carbonyl group against further reaction.

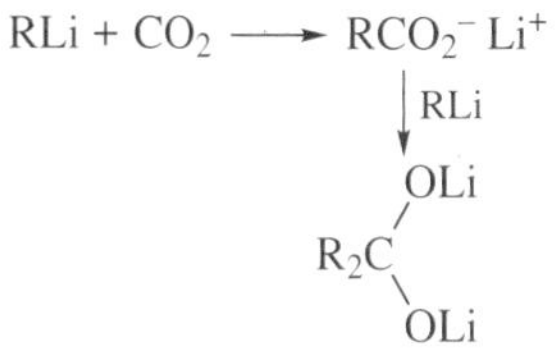

$RLi + CO_2 \longrightarrow RCO_2^- Li^+ \xrightarrow{RLi} R_2C(OLi)_2$

2.2.5 Organocopper and Organocuprate Reagents

The addition of an organometallic reagent to an α,β-unsaturated ketone may take place in a 1,2- or a 1,4-manner. Organocopper (RCu) and organocuprate (R_2CuLi) reagents are the most commonly used compounds for **conjugate (1,4) addition**. The simple organocopper reagents (RCu) are of relatively low reactivity. Although Grignard reagents with a catalytic amount of a copper(I) salt have been used, the much more widely applied reagents are lithium diorganocuprates (R_2CuLi) (Scheme 2.8). The 1,4-addition is particularly favoured in the presence of chlorotrimethylsilane, which will trap an enolate as its trimethylsilyl derivative.

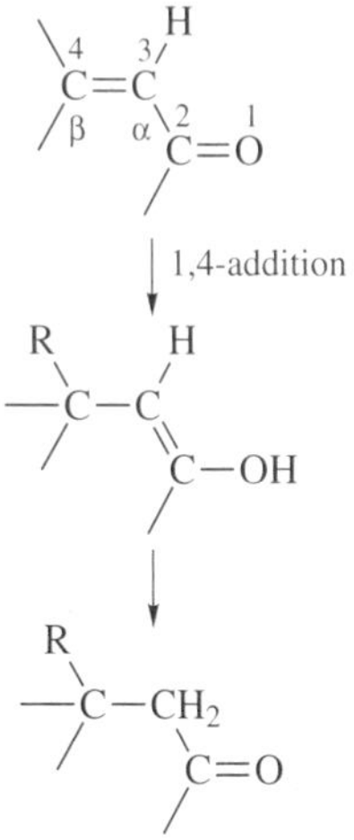

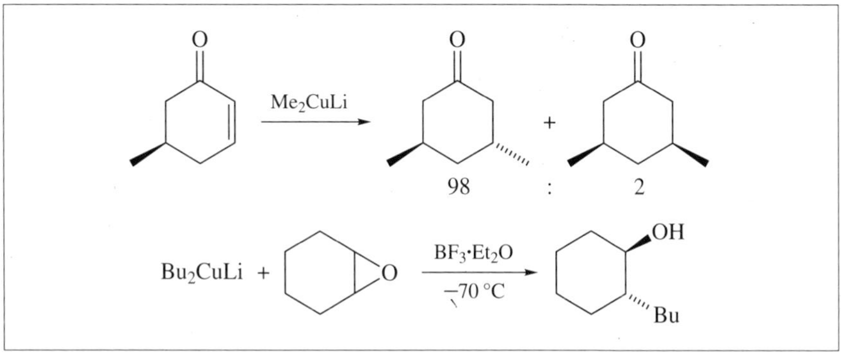

4, β, 3, H, α, 2, 1, C=O; 1,4-addition; R, H, C—OH; R, C—CH₂, C=O

O; Me_2CuLi; O; +; O; 98 : 2

Bu_2CuLi + O; $BF_3{\cdot}Et_2O$, −70 °C; OH, Bu

Scheme 2.8

In terms of hard and soft acids and bases, copper is a typical soft acid. When combined with a hard acid such as boron trifluoride etherate, the reactivity is enhanced. The combined reagent is particularly useful in the cleavage of epoxides (Scheme 2.8).

2.2.6 Organopalladium Compounds

Palladium has played an important role in recent methods of carbon–carbon bond formation. Since palladium is a precious, expensive metal, the most useful reactions employ palladium compounds in catalytic rather than stoichiometric amounts. Palladium has two readily accessible valency states, Pd(0) and Pd(II) [*e.g.* $Pd^0(PPh_3)_4$ and $Pd^{II}Cl_2(PPh_3)_2$]. The former has 10 electrons in the outer valency shell and thus tetracoordinate palladium(0) can achieve a stable 18-electron shell. Tetracoordinate palladium(II) can possess a fairly stable 16-electron outer shell.

$(Ph_3P)_4Pd^0$
Tetrakis(triphenylphosphine)-palladium(0)

Tetrakis(triphenylphosphine)palladium(0) will insert palladium(0) into a carbon–halogen bond in an **oxidative addition** reaction. The palladium(0) is converted to palladium(II). In the general case this may be represented as: Pd(0) + X–Y → X–Pd(II)–Y.

In the reverse of this reaction, known as a **reductive elimination**, a bond between X and Y is formed and the Pd(0) is regenerated. If this can be turned into a cycle during which one of the ligands (X or Y) is replaced by a group Z, then a coupling sequence can be established which is catalytic in palladium (Scheme 2.9) and leads to the formation of YZ.

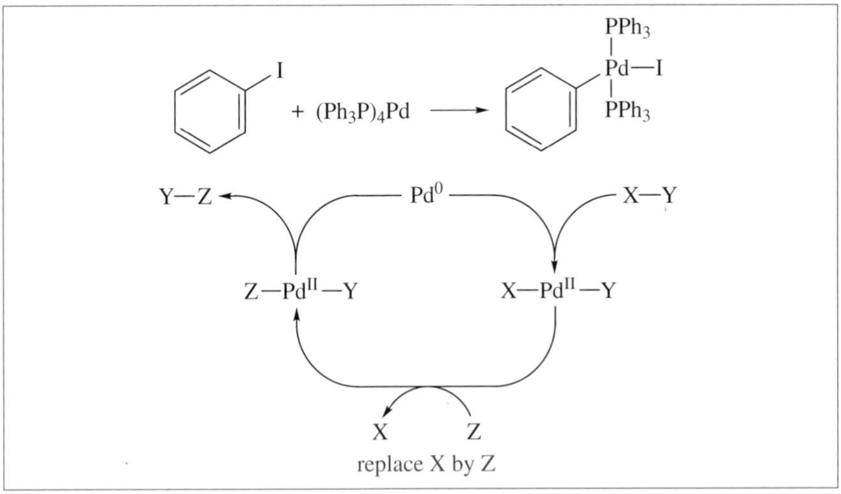

Scheme 2.9

There are a number of useful reactions in which a **ligand replacement** takes place. The use of **carbopalladation** is an important example involving the addition of palladium(0) to an alkene and the subsequent elimination of a β-hydride in the reductive elimination step (Scheme 2.10).

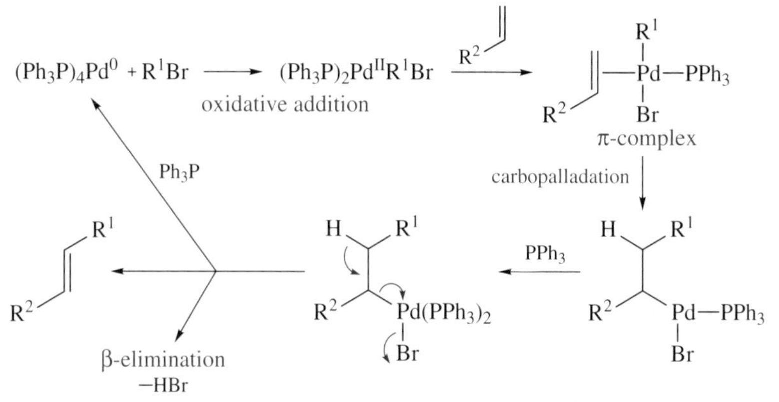

Scheme 2.10

In order to be of value in the regiospecific formation of an alkene, the use of the reaction is restricted to cases in which there is only one β-hydrogen atom which can be readily eliminated. This forms the basis of the **Heck reaction** (Scheme 2.11). The Heck reaction involves the coupling of an arylpalladium species with an alkene and it is particularly useful in attaching a side chain to an aromatic ring.

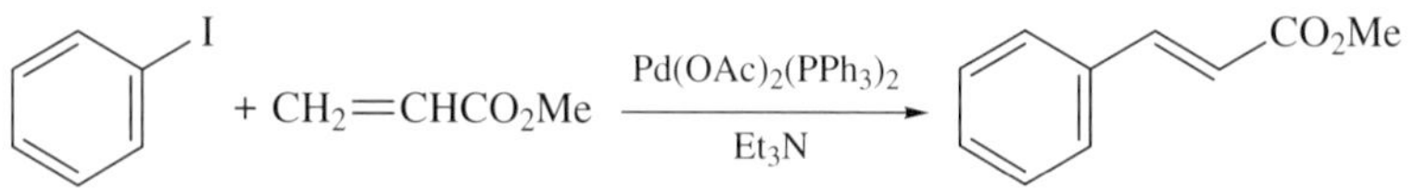

Scheme 2.11 Heck reaction

2.2.7 Cross-coupling Reactions

A number of cross-coupling reactions have been developed using organometallic derivatives of alkenes. The sequence involves an oxidative addition followed by a transmetallation from, for example, a Grignard reagent or a stannane. The **Stille coupling** uses a stannane. For example, a vinylstannane can be coupled with an enol trifluoromethanesulfonate (Scheme 2.12). If more highly substituted alkenes are used, the geometry of the alkene is preserved.

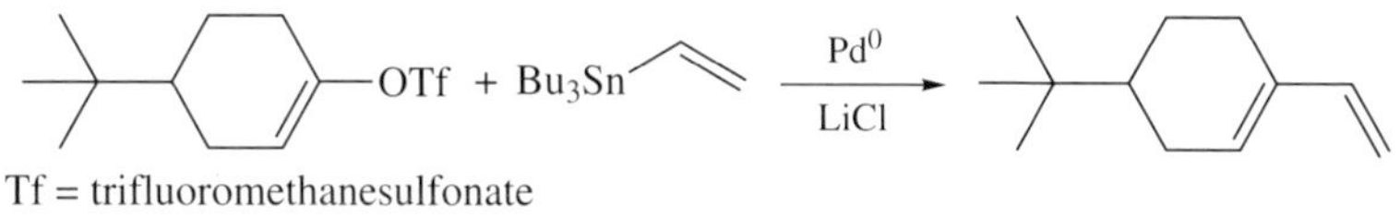

Tf = trifluoromethanesulfonate

Scheme 2.12 Stille coupling

Another cross-coupling reaction is the **Suzuki reaction** which uses a boronic acid. A vinylboronic acid may be prepared by hydroboration of an alkyne (Scheme 2.13).

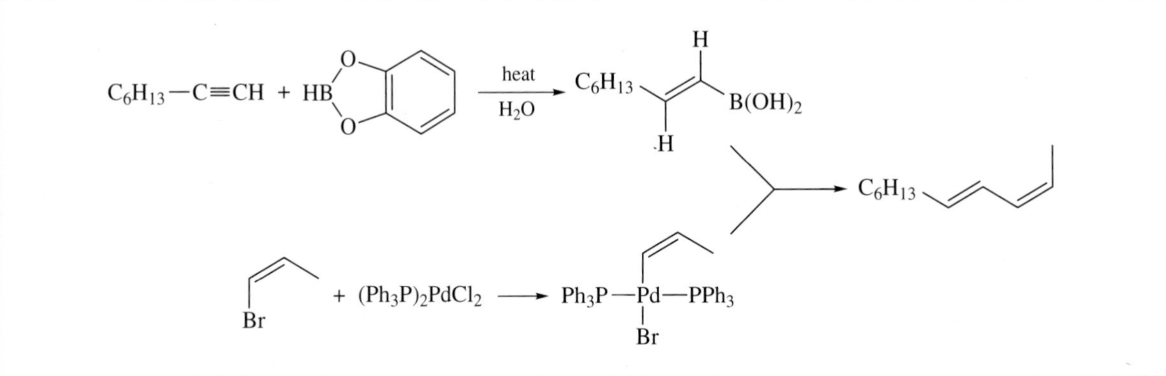

Scheme 2.13 Suzuki reaction

The transmetallation step can involve a copper acetylide in the **Sonogashira coupling** (Scheme 2.14).

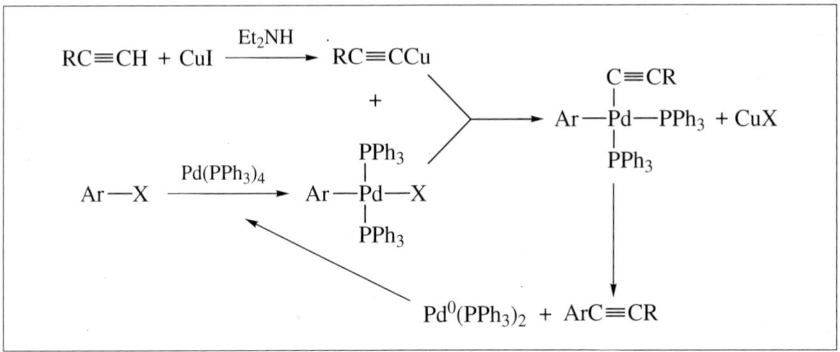

Scheme 2.14 Sonogashira coupling

Complexation of palladium(0) with an alkene activates allylic leaving groups and helps their replacement by another nucleophile. This reaction is discussed later (see Chapter 4).

2.2.8 Orthometallation

The fact that an sp^2 C–H bond is more acidic than an sp^3 C–H underlies the formation and synthetic use of orthometallation in aromatic substitution. When an aromatic ring carries a substituent such as an ether or an amide which can coordinate with the lithium of a lithium alkyl such as butyllithium, the alkyl carbanion can remove a proton from the *ortho* position of the aromatic ring to give an orthometallated lithium aryl. The other product, butane, is volatile. The aromatic carbon is then a powerful nucleophile. It is important to note that this is a reaction of an sp^2 bond, and not of the π-system of the aromatic ring. Thus 1,3-dimethoxybenzene (resorcinol dimethyl ether) can be acylated at the C-2 position (Scheme 2.15), in contrast to the typical electrophilic reaction which takes place at C-4. This illustrates the major value of these reactions in creating adjacent 1,2- rather than 1,4-substitution patterns. The directed orthometallation reaction can be linked to aryl coupling reactions by the use of boronic acid coupling (Suzuki) reactions to form diphenyl derivatives (Scheme 2.15).

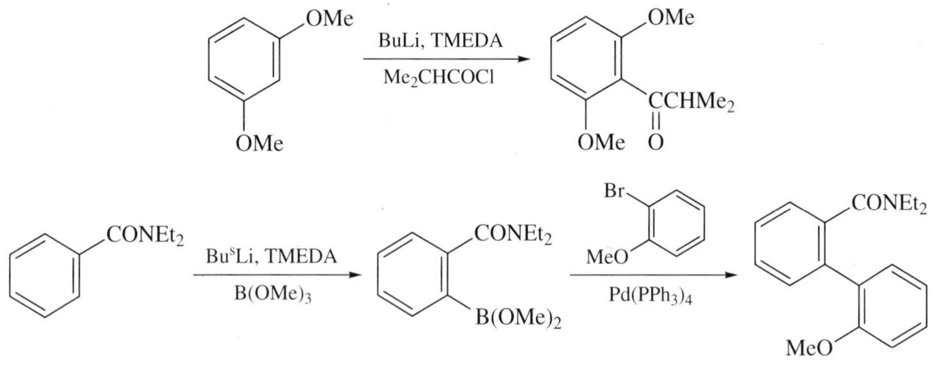

OMe, OMe → (BuLi, TMEDA; $Me_2CHCOCl$) → OMe, $CCHMe_2$ (C=O), OMe

$CONEt_2$ → (Bu^sLi, TMEDA; $B(OMe)_3$) → $CONEt_2$, $B(OMe)_2$ → (Br, MeO aryl; $Pd(PPh_3)_4$) → $CONEt_2$, MeO

Scheme 2.15

Worked Problem 2.2

Q Propose a synthesis of the phthalide **2**.

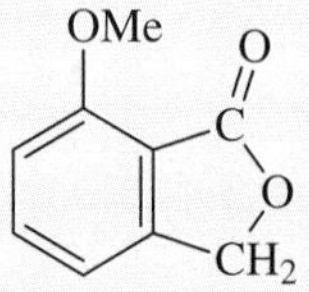

OMe, C=O, O, CH_2 **2**

A The presence of three adjacent substituents is a "marker" for the use of an orthometallation sequence. The methyl ether of salicylic acid (2-hydroxybenzoic acid) is readily available. Although, under the classical conditions of aromatic substitution, a methoxyl group is *ortho/para* directing and a carboxyl group is *meta* directing, orthometallation conditions allow the introduction of a substituent next to the amide of the carboxyl group. The sequence involves lithiation of the diethylamide followed by the introduction of the formyl group and reduction:

This synthesis contrasts with an earlier and longer route which involved the nitration of 3-methoxybenzoic acid to give 3-methoxy-2-nitrobenzoic acid. This was reduced to an amino alcohol, diazotized and converted to a cyanide. Hydrolysis gave the phthalide:

2.3 Acetylides and Nitriles

$RC{\equiv}N \xrightarrow{H_3O^+} RCO_2H$; $RC{\equiv}N \xrightarrow{[H]} RCH_2NH_2$

$RC{\equiv}CR' \xrightarrow{H_3O^+} RCOCH_2R'$; $RC{\equiv}CR' \xrightarrow{[H]} RCH{=}CHR'$

The hydrogen atoms of hydrogen cyanide (H–C≡N) and of a terminal alkyne (H–C≡C–R), in which the carbon is sp hybridized, have sufficient acidity to form carbanions which can be used in substitution and addition reactions to form new carbon–carbon bonds. The value of these two units lies not just in the formation of the new carbon–carbon bonds but also in the subsequent transformations of the cyanide and alkyne. Thus cyanides may be hydrolysed to acids and reduced to amines, while the electron-deficient carbon atom may act as the acceptor for other carbon nucleophiles. Alkynes can be reduced to alkenes and alkanes and converted to methylene ketones.

The substitution and carbonyl addition reactions of sodium acetylide (ethynylsodium) may be used to link components in the synthesis of long-chain natural products such as fatty acids and carotenoids. Examples of the reaction with ketones and the subsequent modification are illustrated by the preparation of an intermediate used in vitamin A synthesis (Scheme 2.16). Another example involves the copper-catalysed oxidative coupling of alkynes (see Chapter 5).

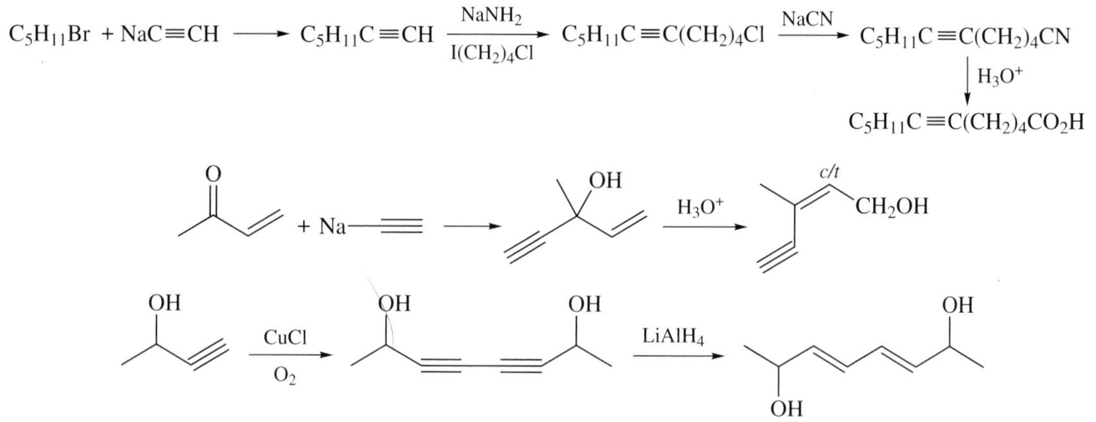

$C_5H_{11}Br$ + NaC≡CH
$C_5H_{11}C≡CH$
$NaNH_2$
$I(CH_2)_4Cl$
$C_5H_{11}C≡C(CH_2)_4Cl$
NaCN
$C_5H_{11}C≡C(CH_2)_4CN$
H_3O^+
$C_5H_{11}C≡C(CH_2)_4CO_2H$
O
+ Na
OH
H_3O^+
c/t
CH_2OH
OH
CuCl
O_2
OH
OH
$LiAlH_4$
OH
OH

Scheme 2.16

The lithium or sodium salt of ethoxyacetylene (ethoxyethyne, EtOC≡CH) will add to carbonyl compounds. The partial hydrogenation of the ethynyl ether to an ethenyl ether followed by hydrolysis gives an aldehyde (Scheme 2.17).

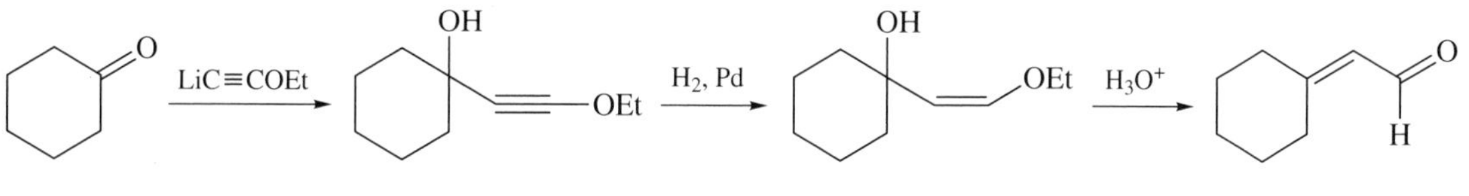

O
LiC≡COEt
OH
OEt
H_2, Pd
OH
OEt
H_3O^+
O
H

Scheme 2.17

Worked Problem 2.3

Q Suggest a synthesis of **3**.

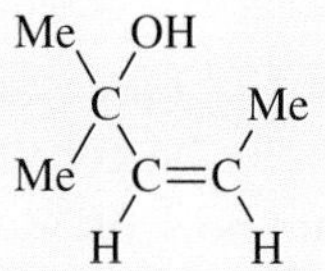

Me
OH
C
Me
Me
C=C
H
H

3

A The presence of a tertiary alcohol suggests an organometallic reaction, and the *cis* double bond in the target molecule might be obtained by the catalytic reduction of an alkyne:

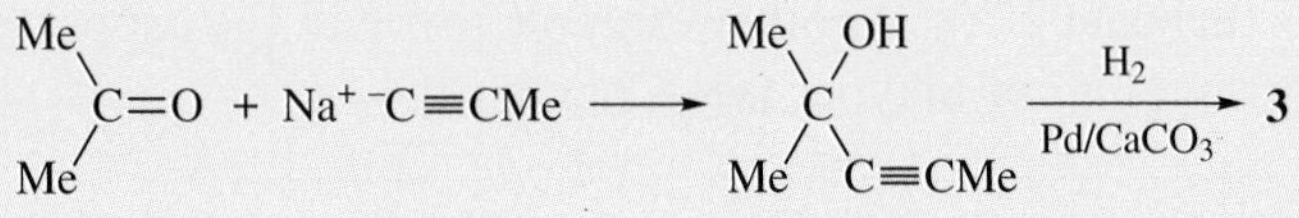

$$\mathrm{Me_2C{=}O} + \mathrm{Na^{+}\,{}^{-}C{\equiv}CMe} \longrightarrow \mathrm{Me_2C(OH)C{\equiv}CMe} \xrightarrow[\mathrm{Pd/CaCO_3}]{\mathrm{H_2}} \mathbf{3}$$

There is an alternative synthesis based on the addition of methylmagnesium iodide to methyl *cis*-but-2-enoate. However, this might be accompanied by problems of 1,4-addition.

2.4 Ylide Reactions

2.4.1 The Wittig Reaction

The elimination of a hydrogen halide from a phosphonium salt by a strong base, such as sodium hydride, leads to the formation of a carbanion. The carbanion is stabilized by the adjacent positively charged phosphorus to form a dipolar substance known as an **ylide**. The nucleophilic carbanion can react with electron-deficient centres such as the carbonyl group. The addition to a carbonyl group generates a dipolar intermediate, a **betaine**, and then a cyclic **1,2-oxaphosphetane**. Decomposition of this adduct, with elimination of a phosphine oxide, brings about the regiospecific formation of an alkene. This reaction is known as the **Wittig reaction** (Scheme 2.18). The reaction is driven by the high negative enthalpy of formation of the phosphorus–oxygen bond.

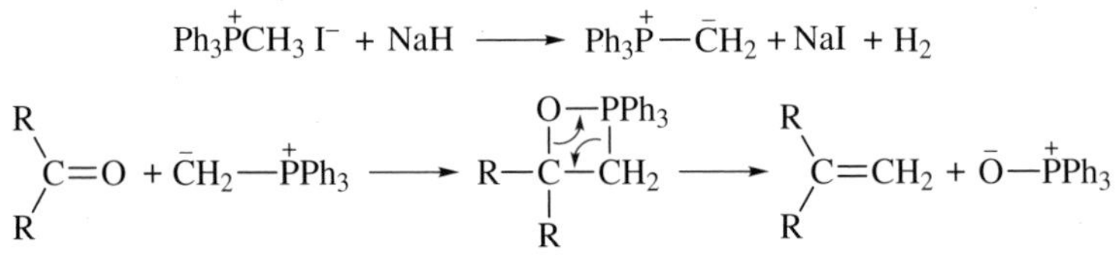

$$\mathrm{Ph_3\overset{+}{P}CH_3\ I^-} + \mathrm{NaH} \longrightarrow \mathrm{Ph_3\overset{+}{P}{-}\overset{-}{C}H_2} + \mathrm{NaI} + \mathrm{H_2}$$

$$\mathrm{R_2C{=}O} + \mathrm{\overset{-}{C}H_2{-}\overset{+}{P}Ph_3} \longrightarrow \mathrm{R{-}\underset{R}{C}(O{-}PPh_3){-}CH_2} \longrightarrow \mathrm{R_2C{=}CH_2} + \mathrm{\overset{-}{O}{-}\overset{+}{P}Ph_3}$$

Scheme 2.18 Wittig reaction

The phosphorus ylide may be obtained from a wide variety of different halides, and thus ketones may be converted into a range of unsaturated compounds. Some examples of these are given in Scheme 2.19. In most cases the phosphonium salts are made by reacting the appropriate alkyl halide with triphenylphosphine.

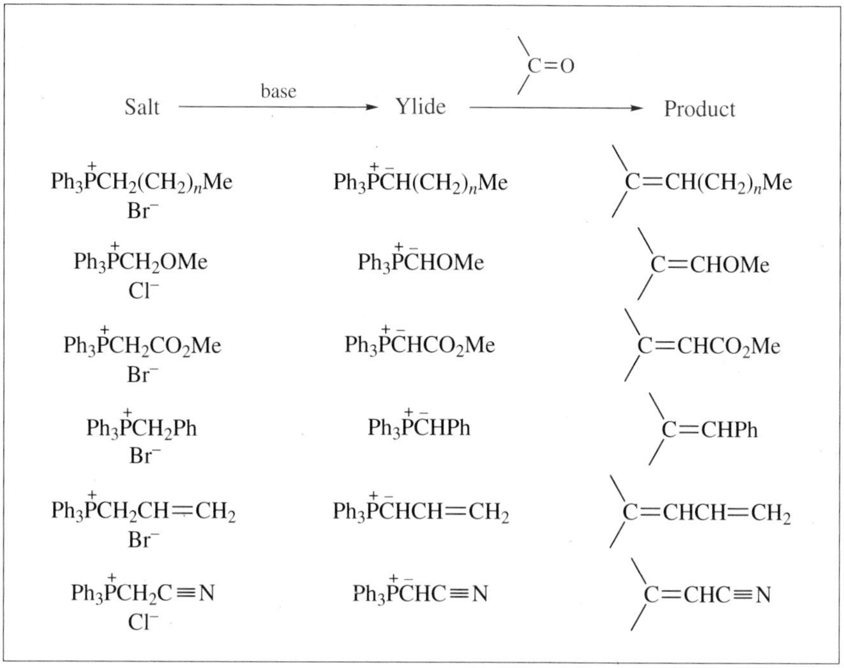

Salt $\xrightarrow{\text{base}}$	Ylide $\xrightarrow{>C=O}$	Product
$Ph_3\overset{+}{P}CH_2(CH_2)_nMe$ Br^-	$Ph_3\overset{+}{P}\overset{-}{C}H(CH_2)_nMe$	$>C=CH(CH_2)_nMe$
$Ph_3\overset{+}{P}CH_2OMe$ Cl^-	$Ph_3\overset{+}{P}\overset{-}{C}HOMe$	$>C=CHOMe$
$Ph_3\overset{+}{P}CH_2CO_2Me$ Br^-	$Ph_3\overset{+}{P}\overset{-}{C}HCO_2Me$	$>C=CHCO_2Me$
$Ph_3\overset{+}{P}CH_2Ph$ Br^-	$Ph_3\overset{+}{P}\overset{-}{C}HPh$	$>C=CHPh$
$Ph_3\overset{+}{P}CH_2CH{=}CH_2$ Br^-	$Ph_3\overset{+}{P}\overset{-}{C}HCH{=}CH_2$	$>C=CHCH{=}CH_2$
$Ph_3\overset{+}{P}CH_2C{\equiv}N$ Cl^-	$Ph_3\overset{+}{P}\overset{-}{C}HC{\equiv}N$	$>C=CHC{\equiv}N$

Scheme 2.19

A useful reaction with the methoxymethylene Wittig reagent leads, via the acid labile enol ether, to an aldehyde. In this way a ketone can be converted to the homologous aldehyde (Scheme 2.20).

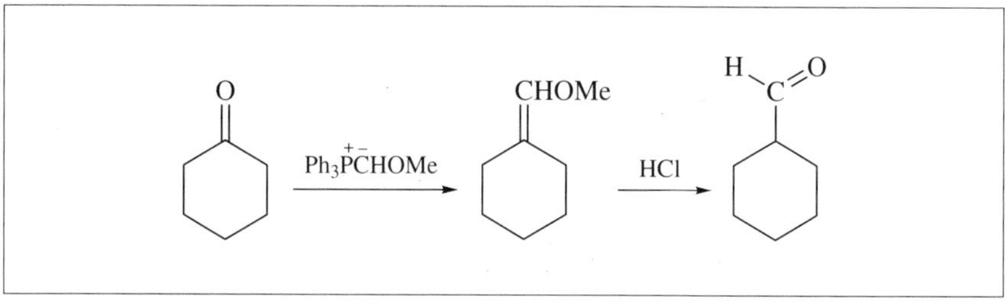

Cyclohexanone (O) $\xrightarrow{Ph_3\overset{+}{P}\overset{-}{C}HOMe}$ (methoxymethylene)cyclohexane (CHOMe) $\xrightarrow{HCl}$ cyclohexanecarbaldehyde (H–C=O)

Scheme 2.20

The **regiospecific** conversion of a carbonyl group to an alkene by the Wittig reaction contrasts with the use of a Grignard reaction and dehydration of the resultant tertiary alcohol, which can lead to an isomeric mixture of alkenes.

Although the Wittig reaction is regiospecific, it is not stereospecific. The geometry of the alkene that is formed is very dependant on the reactivity of the phosphorus ylide and on the presence of metal, particularly lithium, salts. When a reactive ylide interacts with an aldehyde, the geometry of the oxaphosphetane that is formed is determined by the steric approach of the ylide (Scheme 2.21). The reaction tends to give more of the *cis* alkene. If, on the other hand, the Wittig carbanion is stabilized by delocalization of the charge over an adjacent carbonyl group, the first step in the addition is then reversible and the most stable intermediate is formed. Decomposition of the oxaphosphetane yields the *trans* alkene.

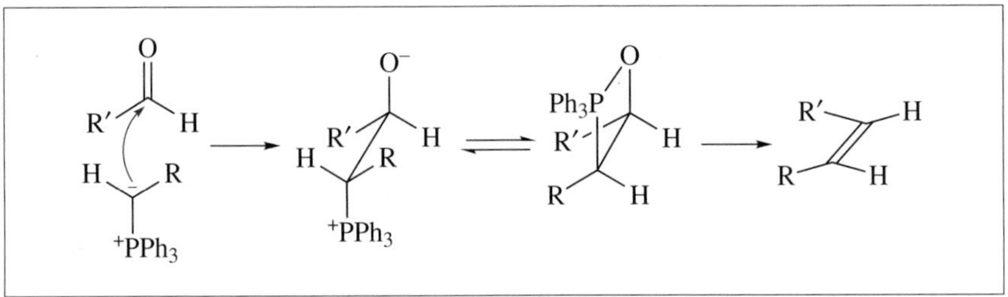

Scheme 2.21

A stabilized ylide may be obtained from a phosphonate in the **Wadsworth–Emmons** variant of the Wittig reaction. When ethyl bromoacetate is heated with triethyl phosphite, a phosphonate is formed. This may behave as a phosphorus analogue of diethyl malonate (diethyl propanedioate). In the presence of a base such as sodium hydride, it gives a stabilized carbanion which can add to a carbonyl group. Decomposition of the intermediate yields an unsaturated ester and a phosphate (Scheme 2.22). It is often difficult to separate the triphenylphosphine oxide from the alkene products of a conventional Wittig reaction. The formation of a phosphate anion in the Wadsworth–Emmons reaction makes removal of the phosphorus easier.

$(EtO)_3P$
Triethyl phosphite
$(EtO)_2P(=O)CH_2CO_2Et$
Triethyl phosphonoacetate [ethyl (diethoxyphosphoryl)acetate]
$CH_2(CO_2Et)_2$
Diethyl malonate

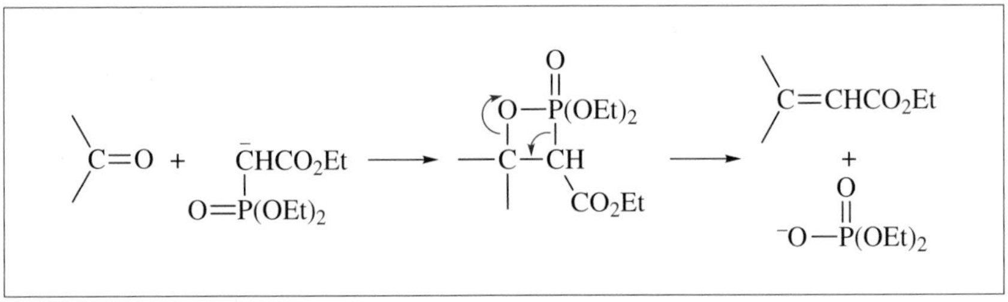

Scheme 2.22
Wadsworth–Emmons reaction

An alternative to the Wittig methylenation of a carbonyl group ($>C=O \rightarrow >C=CH_2$) uses **bis(cyclopentadienyl)dimethyltitanium** (dimethyltitanocene). The reagent may be prepared from bis(cyclopentadienyl)titanium dichloride and methyllithium. It reacts with the carbonyl group to replace the oxygen with a methylene. The other products are methane and bis(cyclopentadienyl)titanium oxide. The reaction is favoured by the ease of formation of the titanium–oxygen bond.

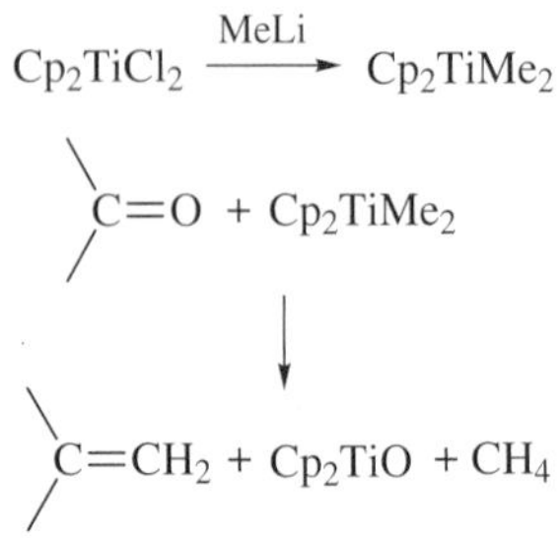

$$Cp_2TiCl_2 \xrightarrow{MeLi} Cp_2TiMe_2$$

$$>C=O + Cp_2TiMe_2 \longrightarrow >C=CH_2 + Cp_2TiO + CH_4$$

Worked Problem 2.4

Q Propose a synthesis of the labelled alkene **4**.

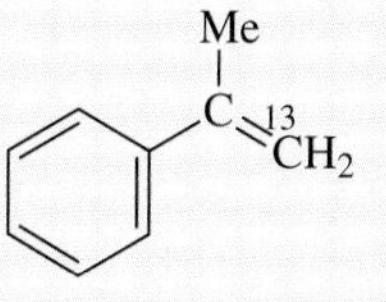

A The $=CH_2$ fragment suggests the use of the Wittig reaction. The synthesis starts from $^{13}CH_3I$. Note that an alternative procedure using a Grignard reaction and dehydration of the resultant alcohol would lead to a scrambling of the label.

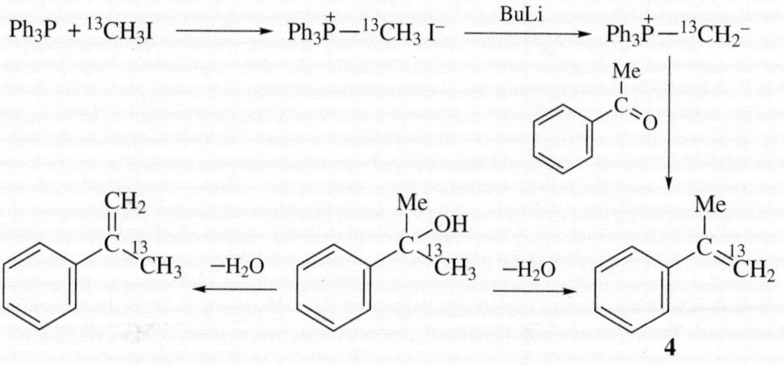

$$Ph_3P + {}^{13}CH_3I \longrightarrow Ph_3\overset{+}{P}{-}{}^{13}CH_3\,I^- \xrightarrow{BuLi} Ph_3\overset{+}{P}{-}{}^{13}CH_2^-$$

2.4.2 Sulfur Ylides

$Me_2\overset{+}{S}{-}\overset{-}{C}H_2$

Dimethylsulfonium methylide

$Me_2\overset{+}{S}(=O){-}\overset{-}{C}H_2$

Dimethylsulfoxonium methylide

The use of sulfur in place of phosphorus brings about a different mode of decomposition of the intermediate betaine. Two sulfur ylides, **dimethylsulfonium methylide** and **dimethylsulfoxonium methylide**, have been used. Both ylides react with ketones to give epoxides, but the stereochemistry of the product may differ (Scheme 2.23).

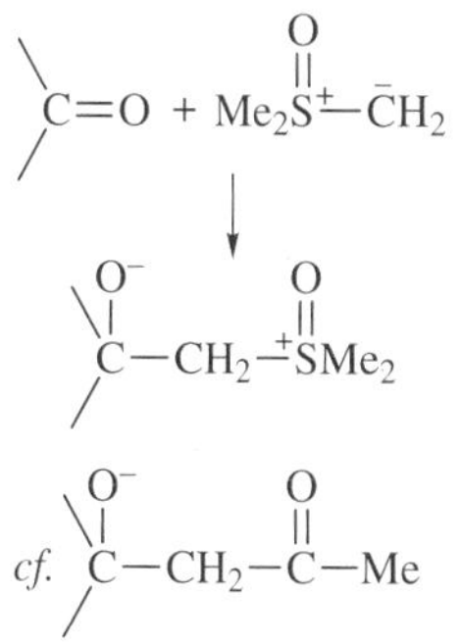

The intermediate arising from the addition of the sulfoxonium ylide resembles the β-hydroxy ketone derived from an aldol condensation. This addition is reversible and hence, by analogy, the addition of the sulfoxonium ylide may be subject to thermodynamic control. In the addition to a cyclic ketone, the carbon atom bearing the sulfoxonium group takes up the equatorial conformation, and the epoxide that is formed possesses an equatorial C–C bond. On the other hand, the addition of a sulfonium ylide is irreversible and kinetic control dominates, leading to axial attack (Scheme 2.24).

$$>C=O + Me_2\overset{+}{S}(=O){-}\overset{-}{C}H_2 \longrightarrow >C(O^-){-}CH_2{-}\overset{+}{S}(=O)Me_2$$

$$cf.\ >C(O^-){-}CH_2{-}C(=O){-}Me$$

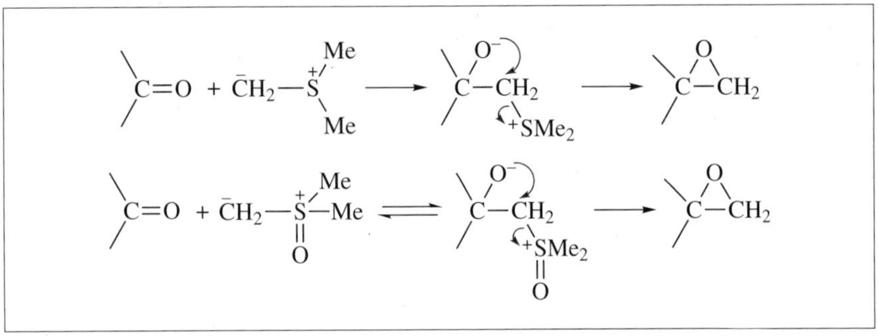

Scheme 2.23

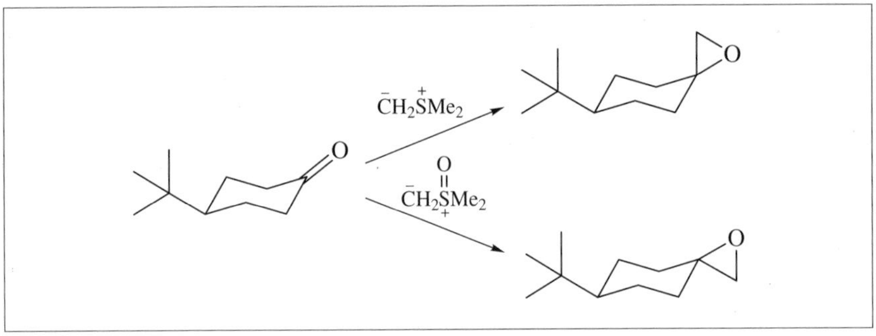

Scheme 2.24

The ability of sulfur to stabilize an adjacent carbanion is not restricted to the sulfonium and sulfoxonium salts. Sulfides, and in particular **1,3-dithioacetals**, can provide synthetically useful stabilized carbanions. Dithioacetals are derived from carbonyl compounds. The propane 1,3-dithioacetal of an aldehyde (a 1,3-dithiane) gives a carbanion which can be used in synthesis. The formation of a carbanion from a 2-substituted 1,3-dithiane derived from an aldehyde represents a reversal of the reactive character of the original aldehyde carbon. It has changed from an electron-deficient carbon to an electron-rich carbon. A synthetic use is shown in Scheme 2.25.

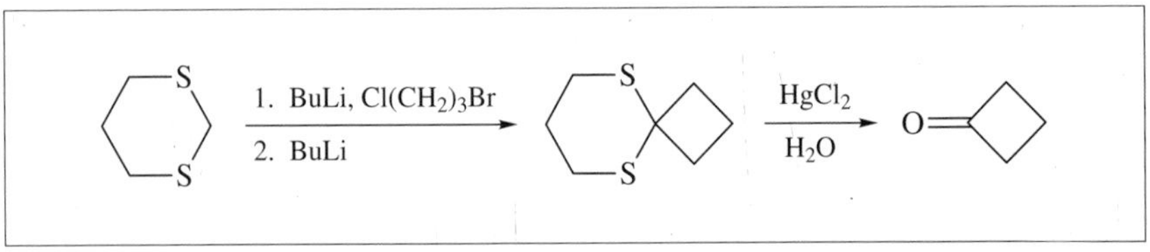

$R_2C(SR)_2$
Dithioacetal

Scheme 2.25

2.5 Silicon and Boron in C–C Bond Formation

2.5.1 Silicon Reagents

Silicon is more electropositive than carbon, and so the carbon–silicon bond is strongly polarized in the sense $Si^{\delta+}$–$C^{\delta-}$. Silicon also has the ability to stabilize an α-carbanion. A further useful property is its ability to stabilize a β-carbocation. This can determine the regioselectivity of a number of reactions between electrophiles and vinyl- or allylsilanes. Finally, many reactions involving organosilicon compounds are helped by the formation of strong silicon–oxygen or even stronger silicon–fluorine bonds at the expense of other weaker bonds.

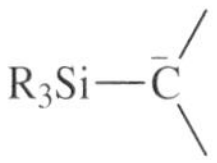

α-Carbanion

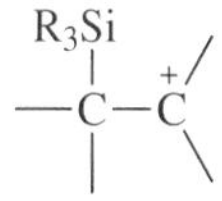

β-Carbocation

The Peterson silicon-based alkene synthesis has some analogy to the Wittig reaction. Although the use of silicon-stabilized carbanions is less common than those derived from phosphorus or sulfur, there are a number of significant advantages in steric terms. The reagent is a trimethylsilylmethyl Grignard which adds to a carbonyl group to form a β-hydroxysilane. Elimination of the trialkylsilyl group and the hydroxyl group then yields the alkene. Which of the two mechanistic pathways this reaction follows depends upon whether the elimination is carried out under acidic or basic conditions (Scheme 2.26). There are different stereochemical consequences with substituted hydroxysilanes. Under acidic conditions an *anti* elimination occurs, whereas under basic conditions a *syn* elimination takes place. Thus the *threo* isomer of 5-(trimethylsilyl)octan-4-ol gives *cis*-octene under acidic (H_2SO_4) conditions and *trans*-octene under basic (NaH) conditions.

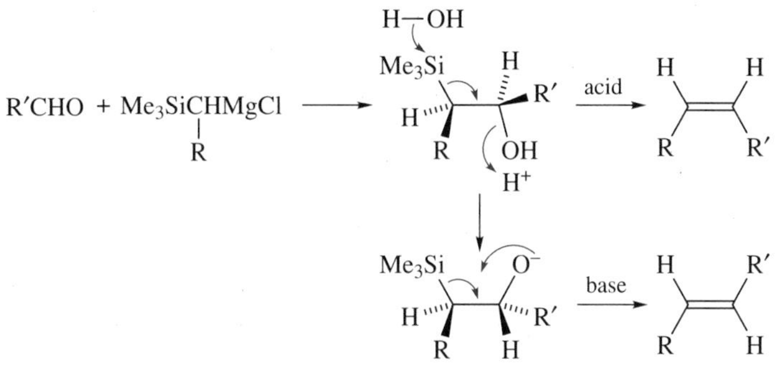

Scheme 2.26 Peterson synthesis

2.5.2 Organoboron Compounds

Boron compounds have three useful properties in a synthetic context. Firstly, the presence of a vacant "p" orbital on the boron means that the boron is a good electrophile and will react with alkenes in the hydroboration reaction. Secondly, boron is able to stabilize an adjacent carbanion. Thirdly, boron readily forms bonds to oxygen. The consequences of this can be seen in the hydroboration reaction (see Chapter 7) and in the carbonylation of boranes In the latter reaction a trialkylborane, formed from the addition of borane to an alkene, adds carbon monoxide (Scheme 2.27). The adduct rearranges to form an acylborane, which in turn undergoes further rearrangement via a borepoxide to a boronic anhydride. This can be oxidatively cleaved by alkaline hydrogen peroxide to form boric acid and a tertiary alcohol. If one of the alkyl groups on the borane is a sterically hindered thexyl (tertiary hexyl) group which is reluctant to undergo the final rearrangement step, the reaction may stop at the borepoxide stage. Oxidative cleavage of this gives a ketone.

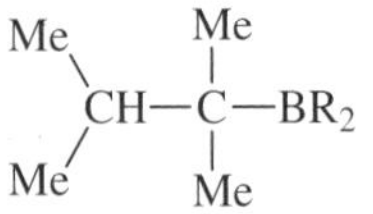

$$Me_2CH{-}CMe_2{-}BR_2$$

Thexylborane

$$R_3B + CO \longrightarrow R_2\overset{-}{B}{-}\overset{+}{C}{=}O \longrightarrow R{-}B(R){-}C({=}O){-}R \longrightarrow \text{borepoxide} \longrightarrow O{=}B{-}CR_3 \xrightarrow[OH^-]{H_2O_2} HO{-}CR_3 + H_3BO_3$$

$$Me_2CH{-}CMe_2{-}BBu_2 \xrightarrow{CO} O{=}CBu_2$$

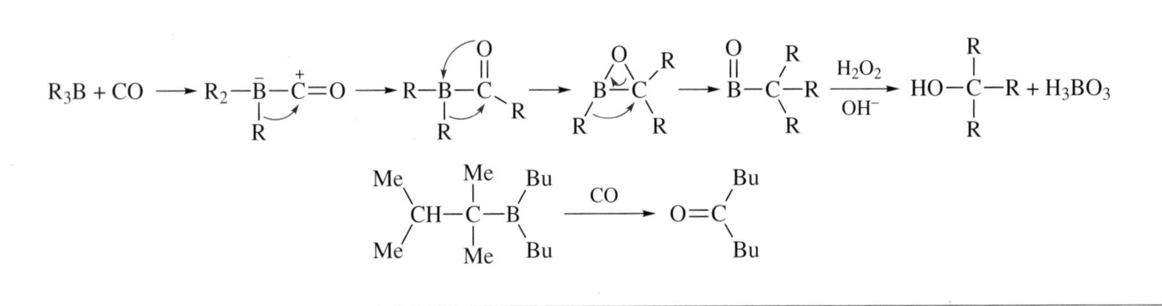

Scheme 2.27

Summary of Key Points

1. Organometallic reagents such as alkylmagnesium halides (Grignard reagents) are prepared from the alkyl halide and the metal and behave as nucleophiles.

2. Addition of the alkyl carbanion to a carbonyl group leads to a hydroxyl group. A "marker" for the Grignard reaction is:

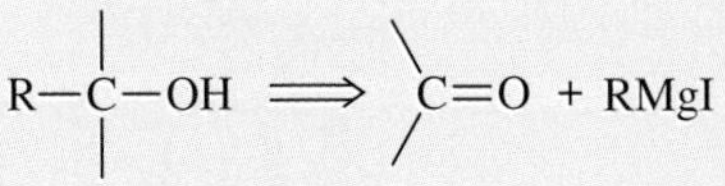

$$R{-}\overset{|}{\underset{|}{C}}{-}OH \Longrightarrow {>}C{=}O + RMgI$$

3. Other metals such as lithium, copper, zinc, tin and palladium form useful organometallic reagents.

4. Coupling reactions based on palladium form a useful strategy for making C–C bonds.

5. The Wittig reaction based on phosphorus ylides is a regiospecific method for converting a carbonyl group to an alkene. The "marker" for a Wittig reaction is:

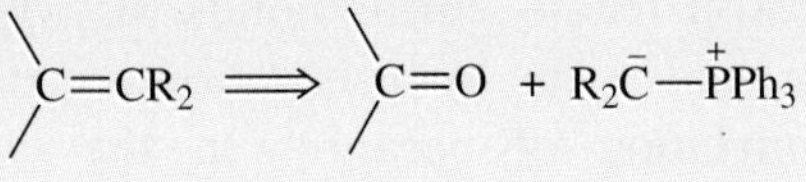

$${>}C{=}CR_2 \Longrightarrow {>}C{=}O + R_2\overset{-}{C}{-}\overset{+}{P}Ph_3$$

6. Compounds of sulfur, silicon and boron can be used to make carbon–carbon bonds.

Problems

2.1. Show how the following labelled compounds might be prepared by using Grignard and related organometallic reagents, together with $^{14}CH_3I$, $H^{14}CHO$ or $^{14}CO_2$ as the source of the label:

(a) $^{14}CH_3CO_2H$ (b) $Me^{14}CO_2H$ (c) $^{14}CH_3CH_2CH_2OH$

(d) $Me^{14}CH_2CH_2OH$ (e) $MeCH_2{}^{14}CH_2OH$ (f) $Me^{14}C(=O)CH_2Me$

2.2. Show how the following compounds might be prepared:

(a) 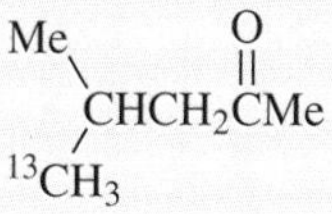 $(Me)(^{13}CH_3)CHCH_2C(=O)Me$ (b) 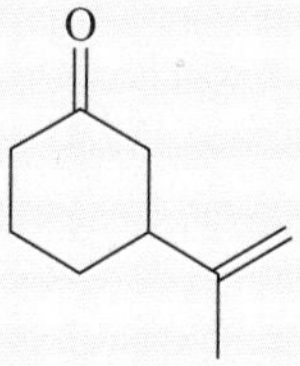 [3-isopropenylcyclohexanone structure]

2.3. Propose a synthetic route for the following compounds:

(a) 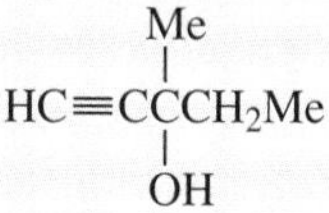 $HC{\equiv}CC(Me)(OH)CH_2Me$ (b) 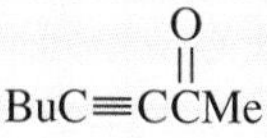 $BuC{\equiv}CC(=O)Me$

2.4. Propose a synthesis of the following compound using $H^{13}C{\equiv}^{13}CH$ as the source of the label:

[5,6-dihydro-2H-pyran-2-one structure, with the C=C carbons labelled 13]

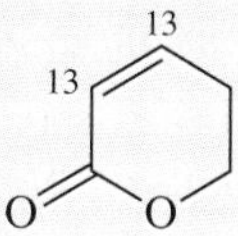

2.5. Show how the following compound might be prepared from cyclohexanone:

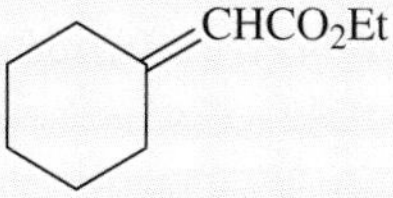

[cyclohexylidene structure: $C_6H_{10}{=}CHCO_2Et$]

2.6. Propose a synthesis of the following compound using a Wittig reaction:

$$MeC(=O)CH{=}C(Me)(CO_2Me)$$

2.7. Propose a synthesis of the following compound using a Wadsworth–Emmons reaction:

$$CH_2{=}C(C(=O)OEt)(H_2C{-}CH{=}CH_2)$$

2.8. Propose a synthesis of the pheromone manicone:

$$Me{-}CH_2{-}CH(Me){-}CH{=}C(Me){-}C(=O){-}CH_2{-}Me$$

Further Reading

P. R. Jenkins, *Organometallic Reagents in Synthesis*, Oxford University Press, Oxford, 1992.

T. L. Ho, *Chemoselectivity in organometallic reactions, a HSAB approach*, in *Tetrahedron*, 1985, **41**, 1.

Y. H. Lai, *Grignard reagents from chemically activated magnesium*, in *Synthesis*, 1981, 165.

D. A. Shirley, *The synthesis of ketones from acid halides and organometallic compounds of magnesium, zinc and cadmium*, in *Org. React.*, 1954, **8**, 28.

A. Krief and A. M. Laval, *Coupling of organic halides with carbonyl compounds promoted by samarium iodide – the Kagan reagent*, in *Chem. Rev.*, 1999, **99**, 745.

G. H. Posner, *Conjugate addition reactions of organocopper reagents*, in *Org. React.*, 1972, **19**, 1; 1975, **22**, 253.

R. J. K. Taylor, *Organocopper conjugate addition–enolate trapping reactions*, in *Synthesis*, 1985, 364.

M. J. Jorgenson. *Preparation of ketones from the reaction of organo-*

lithium reagents with carboxylic acids, in *Org. React.*, 1970, **18**, 1.

J. K. Stille, *The palladium-catalysed cross coupling reactions of organotin reagents with organic electrophiles*, in *Angew. Chem. Int. Ed. Engl.*, 1986, **25**, 508.

R. F. Heck, *The palladium catalysed vinylation of organic halides*, in *Org. React.*, 1982, **27**, 345.

A. De Meijere and F. E. Meyer, *Heck reactions in modern garb*, in *Angew. Chem. Int. Ed. Engl.*, 1994, **33**, 2379.

T. L. Jacobs, *The synthesis of acetylenes*, in *Org. React.*, 1949, **5**, 735.

E. R. H. Jones, *Acetylenes and acetylenic compounds in organic synthesis*, in *J. Chem. Soc.*, 1950, 754.

S. Trippett, *The Wittig reaction*, in *Q. Rev. Chem. Soc.*, 1965, **19**, 1.

A. Maercker, *The Wittig reaction*, in *Org. React.*, 1965, **14,** 270.

J. Reucroft and P. G. Sammes, *Stereoselective and stereospecific synthesis of olefins*, in *Q. Rev. Chem. Soc.*, 1971, **25**, 135.

S. Warren, *Phosphorus and sulfur reagents in synthesis*, in *Chem. Ind. (London)*, 1980, 824.

B. T. Grobel and D. Seebach, *Umpolung of the reactivity of carbonyl compounds through sulfur containing reagents*, in *Synthesis*, 1977, 357.

G. H. Witham, *Organosulfur Chemistry*, Oxford University Press, Oxford, 1995.

I. Fleming, *Some uses of silicon compounds in synthesis*, in *Chem. Soc. Rev.*, 1981, **10**, 83.

D. J. Ager, *The Peterson olefination reaction*, in *Org. React.*, 1990, **38**, 1.

S. E. Thomas, *Organic Synthesis, the Roles of Boron and Silicon*, Oxford University Press, Oxford, 1991.

K. Smith, *Organoboranes as reagents for organic synthesis*, in *Chem. Soc. Rev.*, 1974; **3**, 443.

A. Pelter, *Carbon–carbon bond formation through boron reagents*, in *Chem. Soc. Rev.*, 1982, **11**, 191.

3

Carbonyl Activation and Enolate Chemistry in Carbon–Carbon Bond Formation

Aims

The aim of this chapter is to show how enolate anions and their equivalents can be used in the formation of carbon–carbon bonds. By the end of this chapter you should understand:

- The role of the carbonyl group in rendering an adjacent C–H acidic and in stabilizing the resultant carbanion
- The use of carbanions as nucleophiles in substitution and addition reactions to form C–C bonds
- The application of these principles to other activating groups such as the nitro, nitrile and sulfoxy and phosphono groups

3.1 Introduction

3.1.1 Enols and Enolate Anions

A carbonyl group can make a hydrogen atom attached to an adjacent atom more acidic:

$$-\overset{\overset{\displaystyle O}{\|}}{C}-X-H \rightleftharpoons -\overset{\overset{\displaystyle O}{\|}}{C}-X^- + H^+ \longleftrightarrow -\overset{\overset{\displaystyle O^-}{|}}{C}=X + H^+ \rightleftharpoons -\overset{\overset{\displaystyle OH}{|}}{C}=X$$

In the presence of a base, an anion is formed, which obtains stabilization by delocalization over the carbonyl group. These anionic species behave as nucleophiles and can participate in substitution and addition reactions at electron-deficient centres. The atom adjacent to the carbonyl group may be oxygen as in a carboxylic acid (O=C–OH), nitrogen as in an amide (O=C–NH) or carbon as in a ketone (O=C–CH). The carbanion that is formed from the latter is stabilized in a **resonance** form as an enolate anion. When the nucleophilic carbanion reacts at an electron-deficient carbon, then a new carbon–carbon bond is formed (Scheme 3.1).

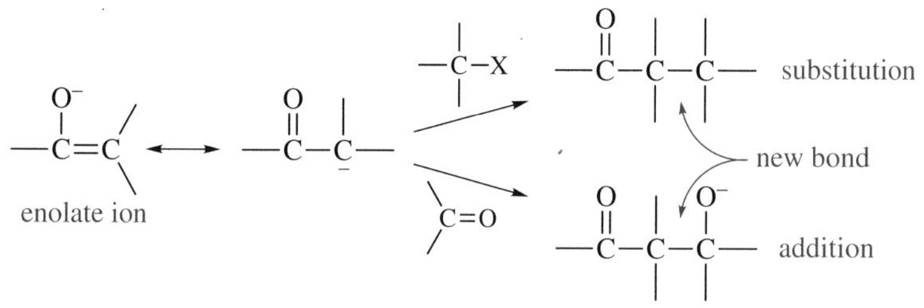

enolate ion

substitution

new bond

addition

Scheme 3.1

It is important to note the position of the new bond that is formed relative to the activating carbonyl group and to contrast this with the position of the new bonds that are formed by organometallic and ylide methods. The relationship between the activating group, the residue of the recipient electron-deficient functional group and the new bond that is formed is a useful **marker** for synthetic strategies based on these reactions. Recognition of this relationship is very helpful in the retrosynthetic analysis of target molecules.

The new bond is formed between the α- and β-carbon atoms:

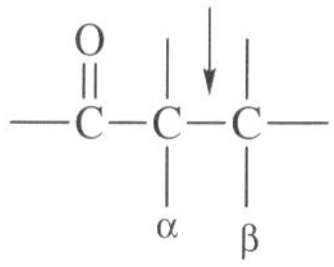

α β

The acidifying effect of a carbonyl group may be modified by substituents. Thus a ketone is a slightly poorer activating group than an aldehyde because of the electron-donating character of the alkyl group. The lone pair of the ester oxygen group further diminishes the activating effect of a carbonyl. Conversion to an anhydride restores the activation since the second carbonyl group "ties up" the lone pairs (Scheme 3.2).

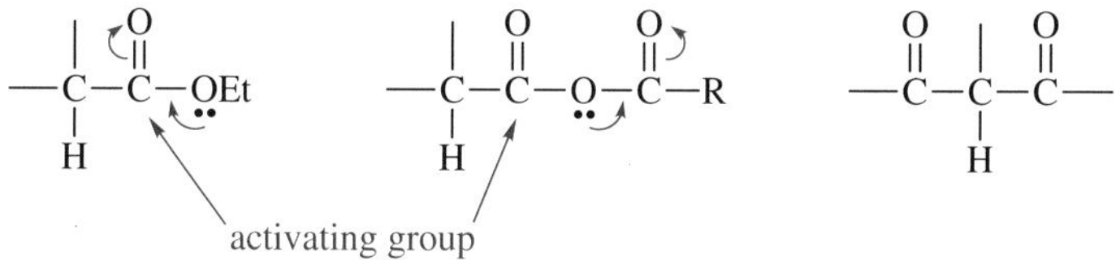

activating group

Scheme 3.2

When two or three carbonyl groups activate the same C–H, the effect is additive and the hydrogen is more acidic. This is reflected in the strengths of the bases that are required to generate various carbanions.

So far we have considered the carbonyl group as the activating group. However, this property is not restricted to the carbonyl group, and there are a number of other synthetically useful activating groups. These have

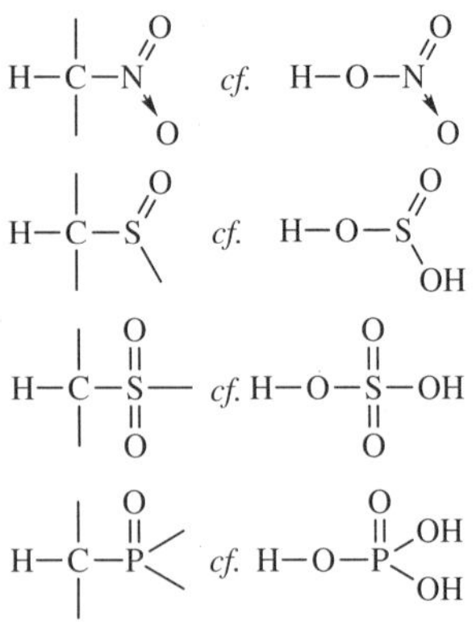

H–C–N(=O)→O *cf.* H–O–N(=O)→O

H–C–S(=O)– *cf.* H–O–S(=O)–OH

H–C–S(=O)₂– *cf.* H–O–S(=O)₂–OH

H–C–P(=O)< *cf.* H–O–P(=O)(OH)–OH

an obvious relationship to well-known oxy acids. Thus the **nitro** group (NO_2) may be compared with nitric acid, the **sulfoxide** (S=O) with sulfurous [sulfuric(IV)] acid, the **sulfone** (SO_2) with sulfuric(VI) acid and the **phosphonate** [$P(O)(OR)_2$] with phosphoric acid.

The oxygen of the activating carbonyl group may also be replaced by another atom such as nitrogen. Thus the **imino** (C=NH) and **nitrile** (C≡N) groups can behave as activating groups. These activating groups may be used synthetically as the sources of carbanions on their own, or in combination with each other or the carbonyl group to doubly activate a C–H and thus generate a carbanion. A list of useful compounds is given in Scheme 3.3.

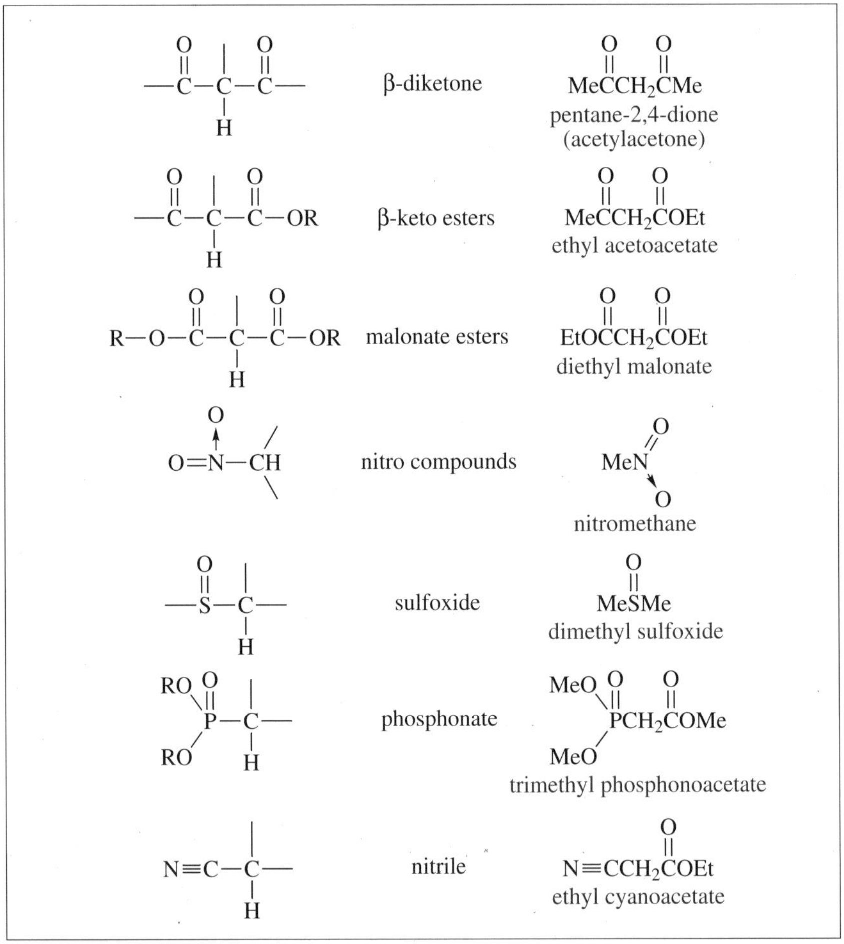

–C(=O)–CH<–C(=O)–	β-diketone	$MeC(=O)CH_2C(=O)Me$ pentane-2,4-dione (acetylacetone)
–C(=O)–CH<–C(=O)–OR	β-keto esters	$MeC(=O)CH_2C(=O)OEt$ ethyl acetoacetate
R–O–C(=O)–CH<–C(=O)–OR	malonate esters	$EtOC(=O)CH_2C(=O)OEt$ diethyl malonate
O=N(→O)–CH<	nitro compounds	$MeNO_2$ nitromethane
–S(=O)–CH<	sulfoxide	MeS(=O)Me dimethyl sulfoxide
(RO)₂P(=O)–CH<	phosphonate	$(MeO)_2P(=O)CH_2C(=O)OMe$ trimethyl phosphonoacetate
N≡C–CH<	nitrile	$N{\equiv}CCH_2C(=O)OEt$ ethyl cyanoacetate

Scheme 3.3

A number of these compounds possess **tautomers**. Thus a carbonyl group possesses a tautomeric relationship with an enol and an aliphatic nitro group has a tautomeric relationship with the *aci* form (Scheme 3.4). The position of the equilibrium in these is important in determining the reactivity of the functional group. For example, many β-formyl ketones exist predominantly as their hydroxymethylene tautomers (Scheme 3.4). When alkylation takes place, it may then occur on carbon or on oxygen.

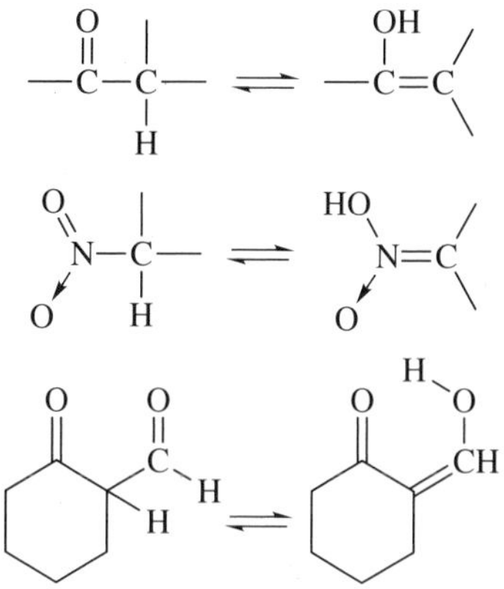

keto ⇌ enol tautomer

nitro compound ⇌ *aci*-nitro compound

β-formyl ketone ⇌ β-hydroxymethylene ketone

Scheme 3.4

3.1.2 Enol Ethers

The relationship with enols provides an alternative way of generating the carbanions (Scheme 3.5). Cleavage of an enol ether, such as a trimethylsilyl enol ether, may be used in the **regiospecific generation of a carbanion**. The value of this in synthesis is that it is possible to separate the enolization step from the carbon–carbon bond formation. Thus where two enols are possible, one may be the rapidly formed kinetic enolate and the other the thermodynamically more stable enolate. Conditions can be devised to favour the formation of one enol ether rather than the other. The enolate anion may then be generated in a second step under conditions in which equilibration cannot take place (*i.e.* aprotic conditions), so that a regiospecific reaction takes place. This is exemplified in Scheme 3.6. The reactions of enol ethers and enamines in forming carbon–carbon bonds are discussed later in this chapter.

An aprotic solvent does not have a hydrogen atom attached to oxygen or nitrogen which can ionize as a proton.

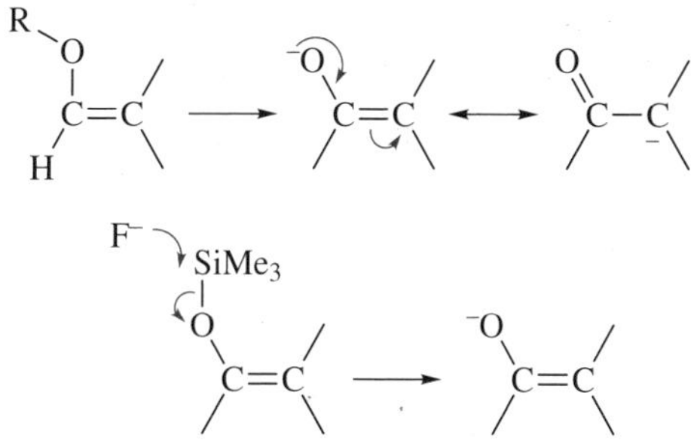

Scheme 3.5

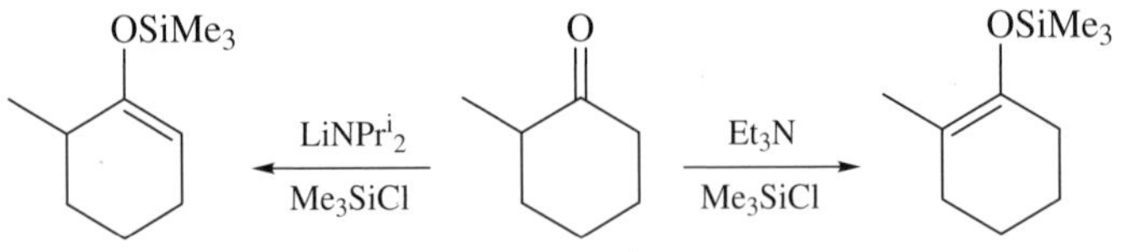

OSiMe$_3$ ← LiNPr$^i_2$ / Me$_3$SiCl — O — Et_3N / Me$_3$SiCl → OSiMe$_3$

Scheme 3.6

Many of the reactions involving carbonyl-activated carbanions are **reversible** and use can be made of this in synthesis. The reversibility of these processes also means that conditions can be found in which the eventual products which are obtained from a reaction are those of thermodynamic control. Hence, although a molecule may possess several potentially reactive centres, conditions may be devised to obtain a single product.

3.2 Alkylation Reactions

$Me_2CHN^- Li^+$
Lithium diisopropylamide
$Me_3C–O^- K^+$
Potassium *t*-butoxide

The use of an enolate anion in the nucleophilic substitution reaction of an alkyl halide, such as iodomethane, leads to an **alkylation reaction**. There are two problems in carrying out an alkylation reaction. The first is to avoid self-condensation reactions in which the enolate anion adds to the carbonyl group of the unionized ketone, and the second is to ensure regiospecificity in the alkylation of unsymmetrical ketones. The first problem may be overcome by using a sufficiently strong base to ensure that all the carbonyl compound is present as the enolate anion. Lithium diisopropylamide (LDA) or potassium *tert*-butoxide are often used in this context. Although these are strong bases, the nitrogen or oxygen atoms are sufficiently sterically hindered to be poor nucleophiles, and thus do not substitute the alkyl halides. One solution to the problem of regiospecificity, by using silyl enol ethers, has already been mentioned. Another solution to the problem involves enhancing the reactivity of a particular C–H by introducing a second auxiliary carbonyl group. Once the alkylation reaction has been carried out, this is removed by making use of the reversibility of carbanion reactions.

β-Keto esters undergo cleavage reactions in two ways. In the presence of acid, the ester group is protonated and hydrolysed to give a β-keto acid, which can then undergo decarboxylation (Scheme 3.7).

Under mildly alkaline conditions the ketonic carbonyl group is more susceptible to nucleophilic addition and a **retro-Claisen reaction** occurs (Scheme 3.7). Hence it is possible to generate methyl ketones or ethyl esters from ethyl acetoacetate (ethyl 3-oxobutanoate, $MeCOCH_2CO_2Et$), depending on the conditions of the cleavage reaction. The presence of an acetyl (MeCO, Ac) group in a target structure can be a good marker for the use of ethyl acetoacetate.

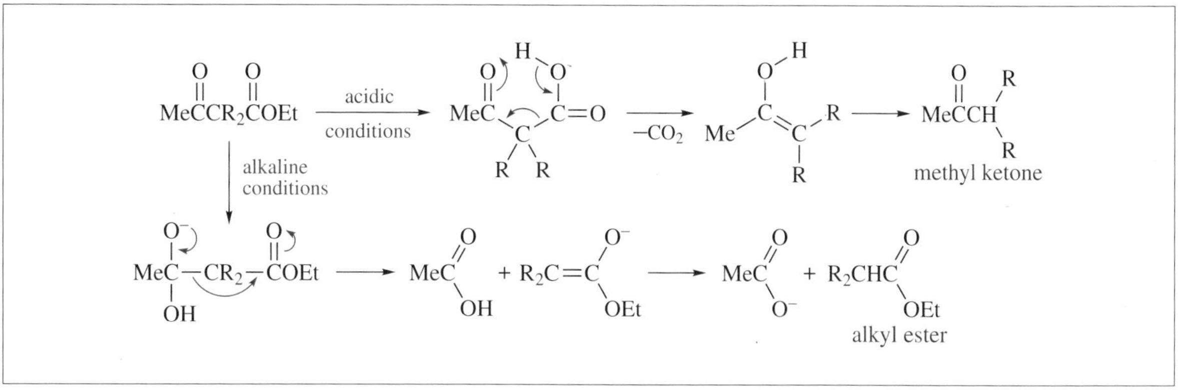

Scheme 3.7

The preparation of 2,2,6-trimethylcyclohexanone and 2,2-dimethylcyclohexanone from 2-methylcyclohexanone illustrates the use of alkylation reactions (Scheme 3.8). In this example, note the way in which the formyl group can be used to activate a centre, and when converted to the sodium salt of its hydroxymethyl tautomer, to separate isomers. The ionic sodium salt is water soluble whilst the non-enolizable dicarbonyl compound is not.

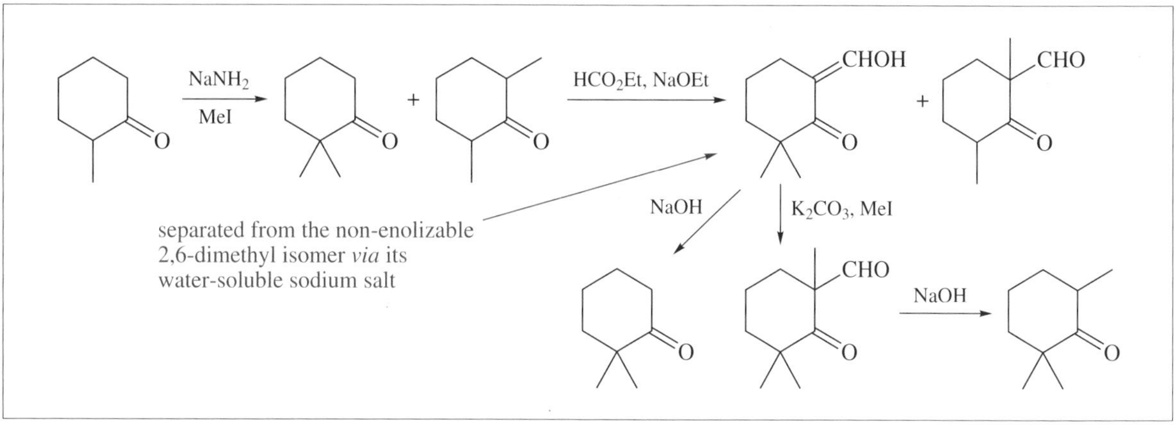

Scheme 3.8

The leaving group in the substitution reaction by the carbanion may be an epoxide. An example of this is the synthesis of acetylcyclopropane (Scheme 3.9). In the first step, ethyl acetoacetate is alkylated by ethylene oxide (1,2-epoxyethane). The oxygen anion from the epoxide participates in an internal hydrolysis of the ester to give acetylbutyrolactone. Hydrolysis of the butyrolactone with hydrogen chloride gives a β-keto acid, which undergoes decarboxylation to form a chloro ketone. This may then be the substrate for an internal alkylation reaction to give acetylcyclopropane.

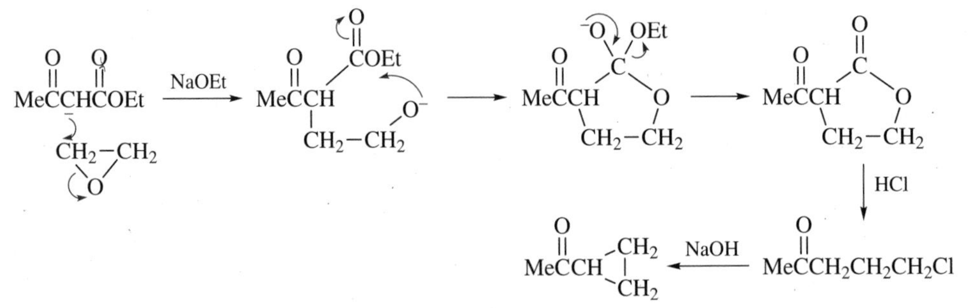

$$\mathrm{MeC(=O)CH(CO_2Et)^{-}} + \mathrm{CH_2{-}CH_2(O)} \xrightarrow{NaOEt} \mathrm{MeC(=O)CH(COEt)CH_2CH_2O^{-}} \longrightarrow \text{tetrahedral intermediate} \longrightarrow \text{lactone} \xrightarrow{HCl} \mathrm{MeCOCH_2CH_2CH_2Cl} \xrightarrow{NaOH} \mathrm{MeCOCH(CH_2)_2}$$

Scheme 3.9

Worked Problem 3.1

In all of the worked problems, note the relationship of the bond that is formed to the activating group.

Q Devise a method for preparing the labelled 3-methylbutan-2-one (1) using $^{14}CH_3I$ as the source of the label.

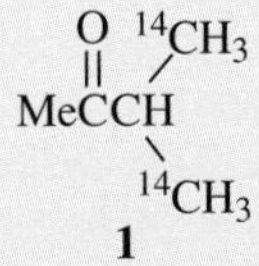

$$\mathrm{MeC(=O)CH(^{14}CH_3)_2} \quad \mathbf{1}$$

A The presence of the labelled methyl groups suggests an alkylation reaction. However, alkylation of propanone would not be suitable because after the first alkylation there is little to differentiate the two enolate anions. Hence there is a need to activate one methylene over the other. This is done by making use of ethyl acetoacetate. After the alkylation, hydrolysis of the ester and decarboxylation of the resultant β-keto acid gives the product. Note the presence of the MeCO group in the target molecule. This is a good "marker" for the use of ethyl acetoacetate in a synthesis.

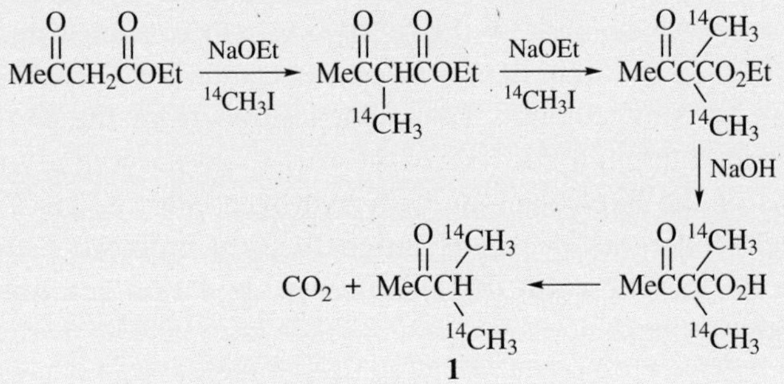

$$\mathrm{MeCOCH_2CO_2Et} \xrightarrow[^{14}CH_3I]{NaOEt} \mathrm{MeCOCH(^{14}CH_3)CO_2Et} \xrightarrow[^{14}CH_3I]{NaOEt} \mathrm{MeCOC(^{14}CH_3)_2CO_2Et} \xrightarrow{NaOH} \mathrm{MeCOC(^{14}CH_3)_2CO_2H} \longrightarrow CO_2 + \mathrm{MeCOCH(^{14}CH_3)_2}\ \mathbf{1}$$

Worked Problem 3.2

Q Propose a carbanion method for the synthesis of 3-methylpentanoic acid (**2**).

$$\underset{\substack{| \\ \mathrm{Me}}}{\mathrm{MeCH_2CHCH_2CO_2H}} \qquad \mathbf{2}$$

A A carbanion dissection would lead to the following:

$$\Longrightarrow \mathrm{MeCH_2\underset{\substack{| \\ Me}}{CH}^+} + {}^-\mathrm{CH_2C(=O)OH}$$

However, the most acidic hydrogen of the carboxylic acid is the O–H, which must be protected as the ester. The acidity of the hydrogen atoms adjacent to the carbonyl group may be increased by the presence of a second activating group. Hence the synthesis involves the alkylation of diethyl malonate (diethyl propanedioate) with 2-bromobutane, followed by hydrolysis of the esters and decarboxylation of the substituted malonic acid:

$$\mathrm{MeCH_2\underset{\substack{| \\ Me}}{CH}Br} + \mathrm{CH_2(CO_2Et)_2} \xrightarrow{\mathrm{NaOEt}} \mathrm{MeCH_2\underset{\substack{| \\ Me}}{CH}CH(CO_2Et)_2}$$

$$\xrightarrow{\mathrm{NaOH}} \mathrm{MeCH_2\underset{\substack{| \\ Me}}{CH}CH(CO_2H)_2} \xrightarrow{\mathrm{HCl}} \underset{\mathbf{2}}{\mathrm{MeCH_2\underset{\substack{| \\ Me}}{CH}CH_2CO_2H}}$$

3.3 Enolate Anions in Carbonyl Addition Reactions

The carbonyl group not only generates a carbanion by making an adjacent hydrogen atom more acidic, but it also provides an electron-deficient centre at which addition reactions can take place. There are three large classes of these reactions, depending on the nature of the carbonyl component. Within each of these classes there are a number of variations, depending on the ultimate fate of the initial addition product. The three major classes are as follows:

1. The **aldol condensation** in which the electron-deficient carbon is an aldehyde or ketone and the product is a β-hydroxy ketone or an α,β-unsaturated ketone (Scheme 3.10).

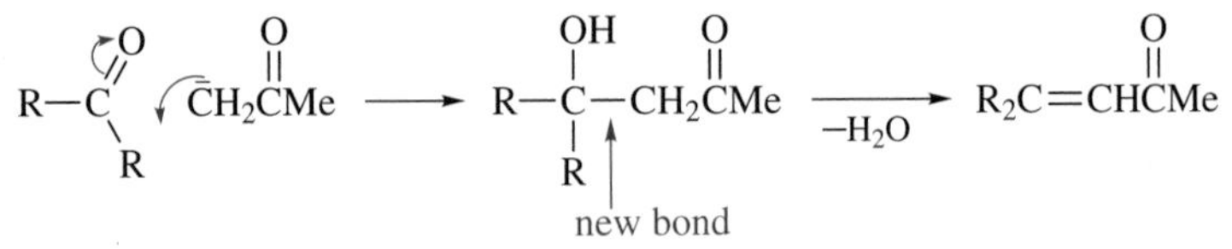

$$\mathrm{R{-}C(=O){-}R} + \mathrm{{}^{-}CH_2C(=O)Me} \longrightarrow \mathrm{R{-}C(OH)(R){-}CH_2C(=O)Me} \xrightarrow{-H_2O} \mathrm{R_2C{=}CHC(=O)Me}$$

new bond

Scheme 3.10 Aldol condensation

2. The **Claisen ester condensation** in which the electron-deficient carbon is an ester carbon and the product is a 1,3-diketone or β-keto ester (Scheme 3.11).

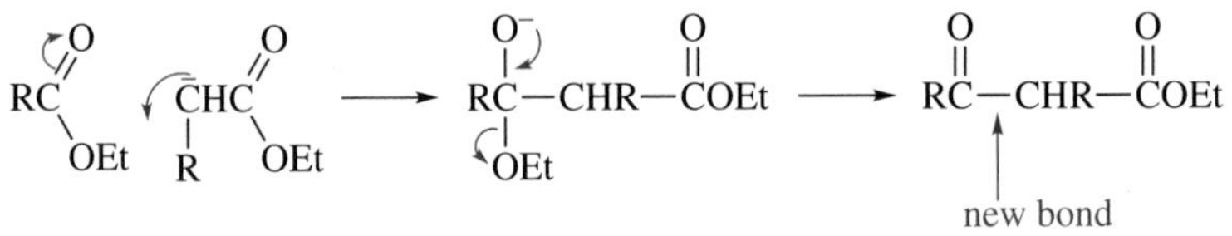

$$\mathrm{RC(=O)OEt} + \mathrm{{}^{-}CHR{-}C(=O)OEt} \longrightarrow \mathrm{RC(O^{-})(OEt){-}CHR{-}C(=O)OEt} \longrightarrow \mathrm{RC(=O){-}CHR{-}C(=O)OEt}$$

new bond

Scheme 3.11 Claisen condensation

3. The **Michael condensation** in which the electron-deficient carbon is the β-carbon of an α,β-unsaturated ketone and the product is a 1,5-diketone (Scheme 3.12).

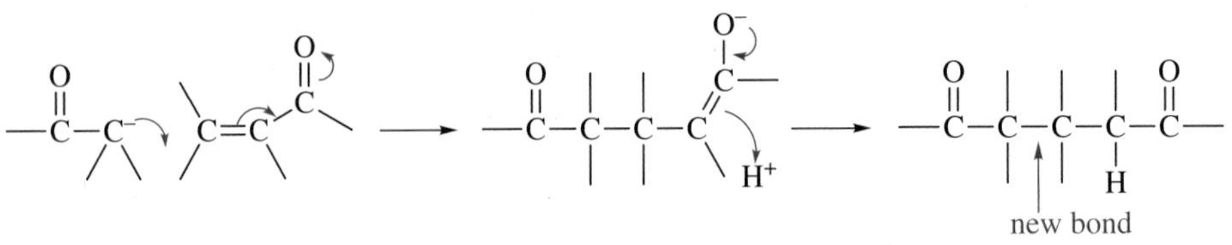

$$\mathrm{{-}C(=O){-}C^{-}\langle} + \mathrm{\rangle C{=}C{-}C(=O){-}} \longrightarrow \mathrm{{-}C(=O){-}C{-}C{-}C{=}C(O^{-}){-}} \xrightarrow{H^+} \mathrm{{-}C(=O){-}C{-}C{-}C(H){-}C(=O){-}}$$

new bond

Scheme 3.12 Michael addition

In each of these it is important to note the position of the new bond that is formed relative to the position of the activating and recipient carbonyl groups. In dissecting a target molecule, note the "**marker**" features of a 1,3-relationship between a carbonyl group and either a hydroxyl group or a second carbonyl group or a transformation product of these. A 1,5-relationship between two carbonyl groups in a target molecule can be a "marker" for the use of a Michael reaction in its synthesis. In designing a synthesis making use of these reactions, it is important to ensure that only the required carbanion is formed and that the electron-deficient centre at which the reaction must occur is the most electrophilic, or at least gives the thermodynamically most stable product.

3.3.1 Aldol Condensations

Aldol condensations may be exemplified by the self-condensation of propanone (acetone) to form 4-hydroxy-4-methylpentan-2-one (diacetone alcohol) and its dehydration product, 4-methylpent-3-en-2-one (mesityl oxide). Note the formation of the unsaturated ketone in this reaction (Scheme 3.13).

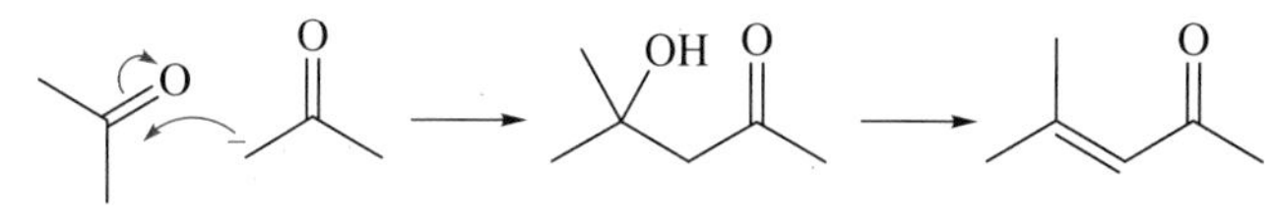

Scheme 3.13

Examples of the range of condensation products that may be obtained from the reaction of different carbanions with an aromatic aldehyde, benzaldehyde, are shown in Scheme 3.14. Many of the reactions are quite general and are known by the names of the chemists who discovered or developed them. They can be distinguished by the different modes of collapse of the intermediate adduct. The presence of an aromatic ring often leads to dehydration of the adduct and the formation of an alkene in conjugation with the aromatic ring.

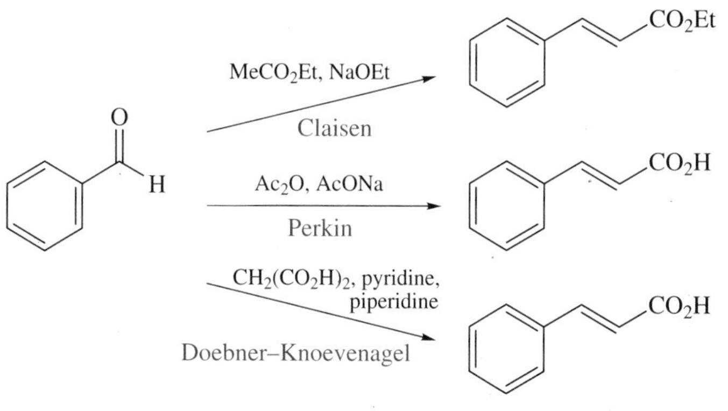

$MeCO_2Et$, NaOEt — Claisen — CO_2Et

Ac_2O, AcONa — Perkin — CO_2H

$CH_2(CO_2H)_2$, pyridine, piperidine — Doebner–Knoevenagel — CO_2H

Scheme 3.14 Claisen, Perkin and Doebner–Knoevenagel condensations

When malonic acid (propanedioic acid) is the source of the carbanion, a decarboxylation occurs with the formation of cinnamic acid (3-phenylpropenoic acid) (the **Doebner–Knoevenagel reaction**). In the case of a succinate (butanedioate) ester, the oxygen anion of the adduct may act as an internal nucleophile and facilitates the hydrolysis of one of the ester groups (the **Stobbé reaction**, Scheme 3.15). The displacement of a chlorine atom and the formation of an epoxide takes place in the **Darzens condensation** with ethyl chloroacetate. The epoxide which is formed (known as a glycidic ester) may be used in a further reaction to generate an aldehyde (Scheme 3.16). The eventual outcome is to extend the carbon chain by one carbon atom.

Scheme 3.15 Stobbé reaction

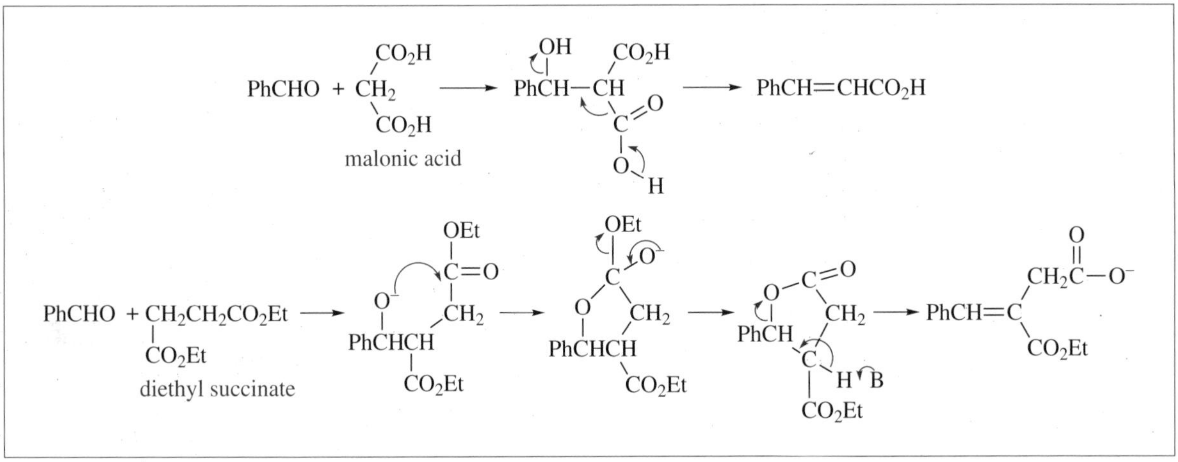

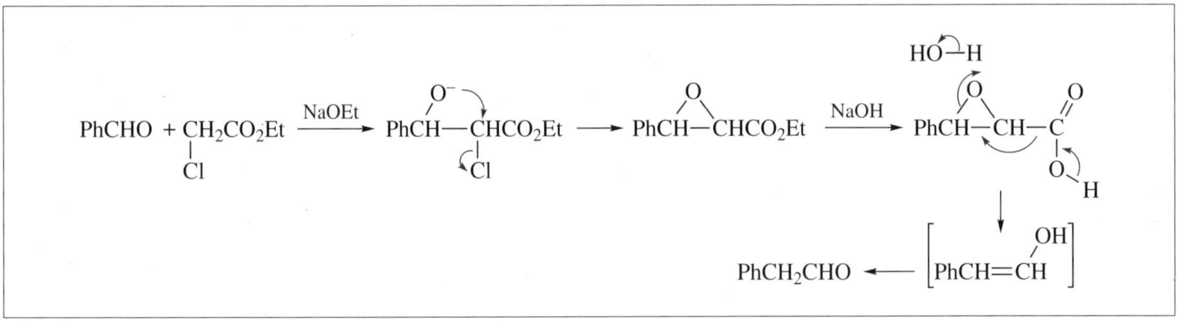

Scheme 3.16 Darzens condensation

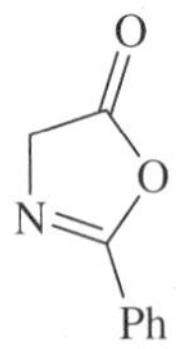

2-phenyl-1,3-oxazol-5(4*H*)-one

The addition of nitromethane yields, after the loss of water, an ω-nitrostyrene (Scheme 3.17). The reduction of these gives rise to β-phenylethylamines, many of which are biologically active. The reaction with the oxazolone (azlactone) derived from benzoylglycine (hippuric acid) can lead to amino acids and is discussed later (see Chapter 6) and the use of phosphonates in the Wadsworth–Emmons reaction was described in Chapter 2.

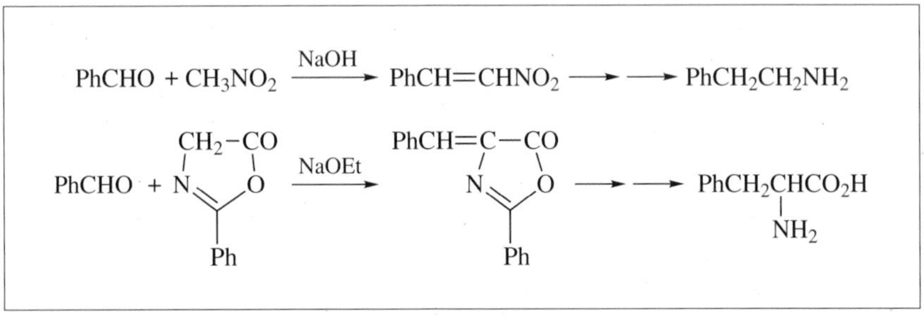

Scheme 3.17

3.3.2 Claisen Condensations

The Claisen condensation of an enolate carbanion with an ester to give a **1,3-diketone** has widespread application. It may involve the self-condensation of two molecules of an ester, as in the formation of ethyl acetoacetate from ethyl acetate (ethyl ethanoate) (Scheme 3.18).

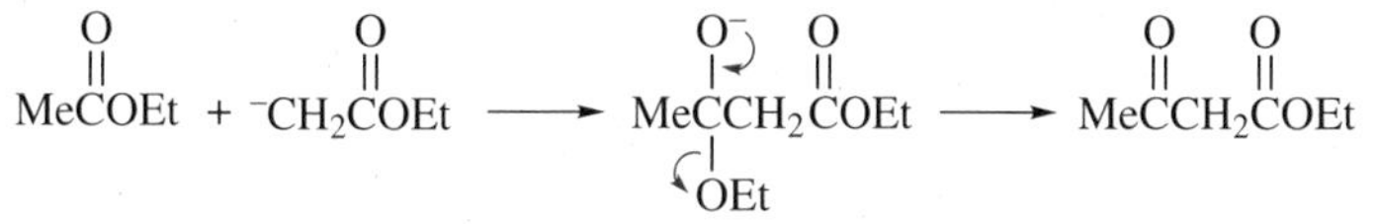

MeC(=O)OEt + $^{-}CH_2C(=O)OEt$ ⟶ $MeC(O^{-})(OEt)CH_2C(=O)OEt$ ⟶ $MeC(=O)CH_2C(=O)OEt$

Scheme 3.18

Where two unlike components are involved, the carbanion must be formed preferentially from one component; the other component, providing the electron-deficient centre, should not be readily enolizable. Some typical non-enolizable esters that are used in this context are diethyl oxalate (ethanedioate), ethyl formate (methanoate), diethyl carbonate and methyl benzoate. Various methods for overcoming this restriction have been based on the acylation of enol ethers and enamines under acid-catalysed conditions.

CO_2Et–CO_2Et
diethyl oxalate

HC(=O)OEt
ethyl formate

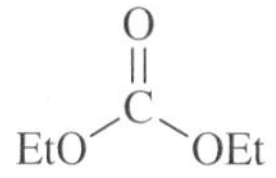

EtO–C(=O)–OEt
diethyl carbonate

PhC(=O)OMe
methyl benzoate

The Claisen condensation of diesters to form rings is known as the **Dieckman cyclization**. This forms a useful methods for making cyclopentanones and cyclohexanones (Scheme 3.19). The cyclization products are stabilized as their enolates, which are converted to the β-keto esters on protonation. These enolates may also be alkylated. Small rings are not formed.

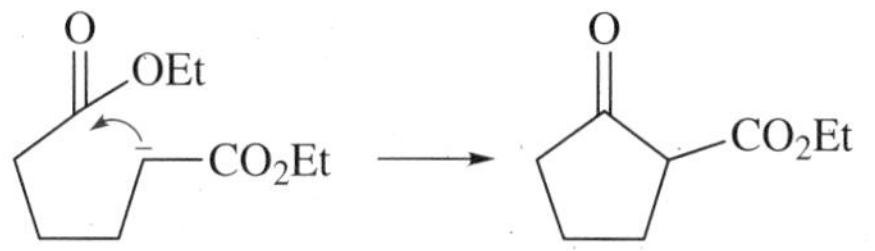

Scheme 3.19 Dieckmann cyclization

Dinitriles may be used in the same way to give β-imino nitriles, which may be hydrolsed to β-keto acids and their decarboxylation products. Although the simple sequence suffers from the same restrictions on ring size, a high dilution technique (the **Thorpe–Ziegler reaction**) has been developed which can be used for the formation of medium and larger rings.

Worked Problem 3.3

Q Suggest a preparation of **3**:

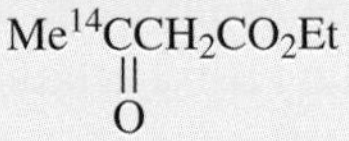

$$\underset{\underset{\displaystyle O}{\|}}{Me^{14}C}CH_2CO_2Et \quad \mathbf{3}$$

A Ethyl acetoacetate is normally prepared by a Claisen condensation of two molecules of ethyl acetate:

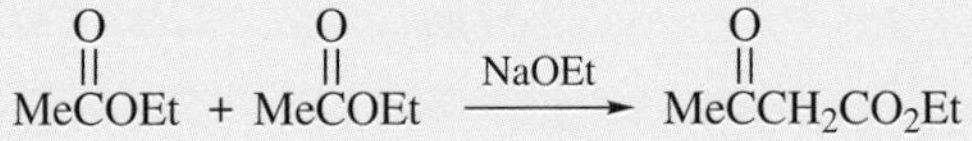

$$MeC(=O)OEt + MeC(=O)OEt \xrightarrow{NaOEt} MeC(=O)CH_2CO_2Et$$

However, in this case the acetate units must be distinct and the regiospecificity of the reaction must be designed to produce the label at C-3 and not at C-1. Hence both the acidity of one C–H and the electron deficiency of the carbonyl group bearing the label must be increased. The synthesis involved preparing [1-^{14}C]acetyl chloride and reacting this with a malonate carbanion. In practice, the mixed *t*-butyl/ethyl ester of malonic acid was used, because the *t*-butyl ester could be removed selectively under acidic conditions:

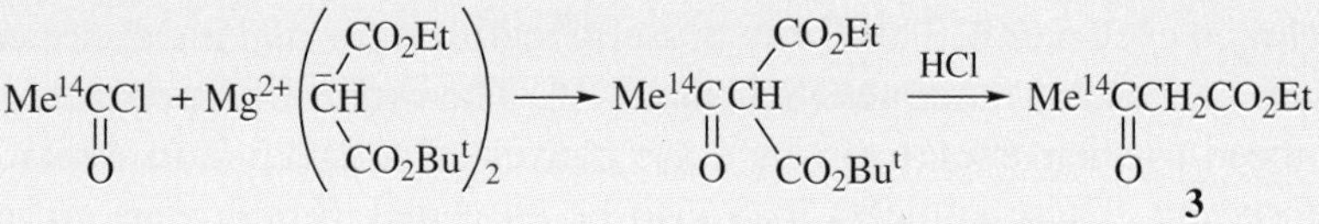

$$Me^{14}C(=O)Cl + Mg^{2+}\left(\bar{C}H(CO_2Et)(CO_2Bu^t)\right)_2 \longrightarrow Me^{14}C(=O)CH(CO_2Et)(CO_2Bu^t) \xrightarrow{HCl} Me^{14}C(=O)CH_2CO_2Et \quad \mathbf{3}$$

Worked Problem 3.4

Q Propose a synthesis of **4**:

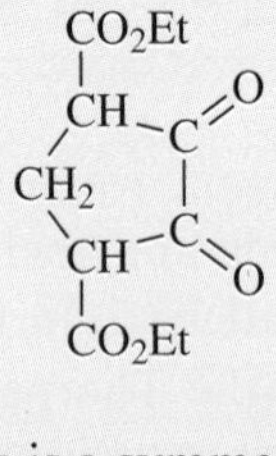

$$\mathbf{4}: \text{ cyclic } CH(CO_2Et)-CH_2-CH(CO_2Et)-C(=O)-C(=O)-$$

A This cyclopentanedione is a symmetrical bis-β-keto ester. β-Keto esters are prepared by Claisen condensations. A retro-Claisen dissection of the structure suggests a condensation between diethyl glutarate (diethyl pentanedioate) and diethyl oxalate:

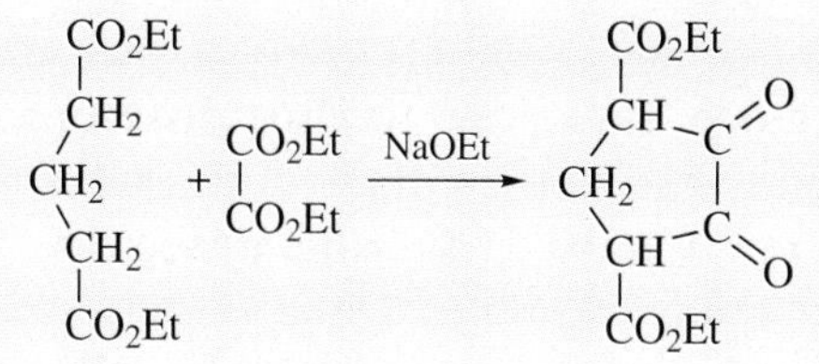

$$\begin{array}{c} CO_2Et \\ | \\ CH_2 \\ / \\ CH_2 \\ \backslash \\ CH_2 \\ | \\ CO_2Et \end{array} + \begin{array}{c} CO_2Et \\ | \\ CO_2Et \end{array} \xrightarrow{NaOEt} \text{cyclic diketone with } CH(CO_2Et), CH_2, CH(CO_2Et), C{=}O, C{=}O$$

3.3.3 The Michael Addition

The Michael reaction involves the addition of a carbanion to the β-position of an α,β-unsaturated ketone. The addition of the anion from diethyl malonate to diethyl fumarate (*trans*-butenedioate) (Scheme 3.20), to form tetraethyl propane-1,1,2,3-tetracarboxylate, is an example of a Michael addition. The Michael reaction may also be used in building ring systems. A sequence known as the **Robinson ring extension** is based on the Michael addition of a cyclohexanone anion to methyl vinyl ketone (3-oxobut-1-ene), followed by an internal aldol condensation (Scheme 3.20). This sequence has been widely used in steroid and terpenoid syntheses.

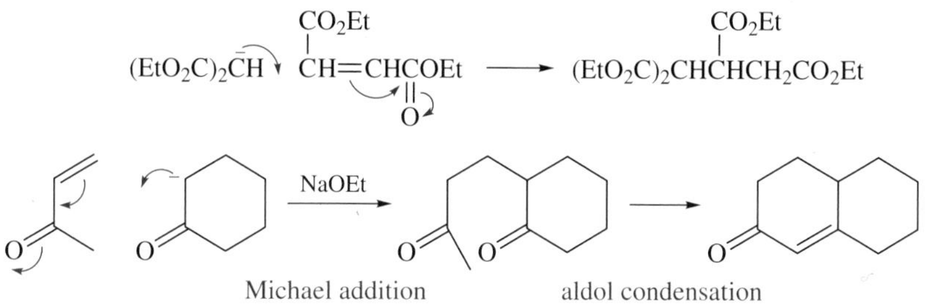

$$(EtO_2C)_2\bar{C}H \;\; \overset{CO_2Et}{\overset{|}{C}H}{=}CH\overset{\overset{}{}}{C}OEt(=O) \longrightarrow (EtO_2C)_2CH\overset{CO_2Et}{\overset{|}{C}H}CH_2CO_2Et$$

NaOEt

Michael addition

aldol condensation

Scheme 3.20

Worked Problem 3.5

Q Suggest a synthesis of dimedone, 5,5-dimethylcyclohexane-1,3-dione (**5**).

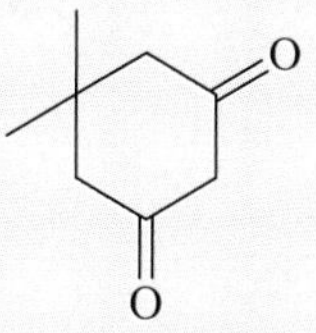

5

A Analysis of the structure of dimedone reveals both a 1,5- and a 1,3-relationship between the carbonyl groups. A 1,3-relationship implies a Claisen condensation between an ester and a carbanion adjacent to a ketone, suggesting the first dissection. The 1,5-relationship between the carbonyl groups in the sub-structure implies a Michael addition, leading to a further dissection. The α,β-unsaturated ketone 4-methylpent-3-en-2-one (mesityl oxide), which is then revealed, is the product of an aldol condensation between two molecules of propanone (acetone). Having carried out a dissection, it is important to consider how this synthesis might be realized in practice. In a mixture of ethyl acetate and 4-methylpent-3-en-2-one the C–H bonds of the acetate are less acidic than those activated by the ketone in the pentenone. Consequently, the acidity of the acetate C–H bonds has to be increased by using diethyl malonate. Once the addition and cyclization have been carried out, the remaining ester group is hydrolysed and the β-keto acid decarboxylated. In this example it is also worth noting that although other carbanion cyclizations are possible, they lead to the formation of four-membered rings. There is an alternative strategy in which the 1,5-dissection is made first. However, this would give a more problematic β-diketone synthesis.

Me Me O O 1:3 ⟹ 1:5 Me Me CH_2 Me O O OEt ⟹ H Me Me C C C O Me $^-CH_2$ C—OEt O

⇓

1.3

Me—C(=O)—Me + $^-CH_2C(=O)Me$ ⟸ Me Me C(OH) CH_2 C(=O) Me

Me₂C=CH–C(=O)Me + $EtO_2C–CH_2–CO_2Et$ —NaOEt→ Me₂C–CH_2–C(=O)–CH_2–C(=O)–$CH(CO_2Et)$ —1. KOH, 2. HCl→ 5,5-dimethylcyclohexane-1,3-dione

3.3.4 Cyclizations

In determining whether a carbanion cyclization is a realistic possibility, it is important to consider whether the transition state can be reached without a serious distortion of the normal bond angles or bonding distances. **Baldwin's rules** describe the relative ease with which carbanion cyclization reactions involving first-row elements occur. In these rules, cyclizations are classified in terms of three criteria. The first criterion is the size of the ring that is being formed and reflects the influence of ring strain. The second is the direction (exo- or endocyclic) in which the bond to the electron-deficient component points relative to the new ring. This reflects the direction of approach of the carbanion. The third criterion is the hybridization (tetrahedral, trigonal or digonal) of the electron-deficient component. This reflects the influence of the transition state geometry on the reaction. Thus the cyclizations in Scheme 3.21 which form six-membered rings are known as 6-*exo*-trig and 6-*endo*-trig, respectively.

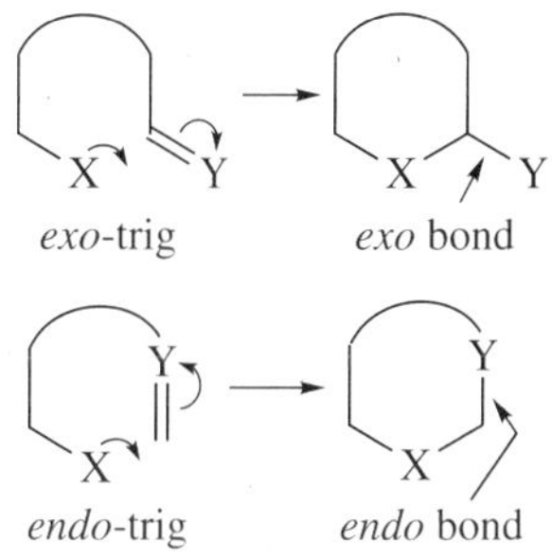

exo-trig *exo* bond

endo-trig *endo* bond

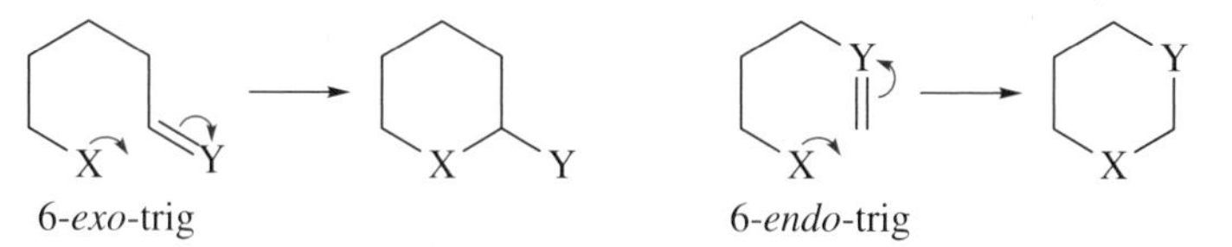

6-*exo*-trig 6-*endo*-trig

Scheme 3.21

The rules can be summarized as follows:

System	exo *bonds to electron-deficient atom*	endo *bonds to electron-deficient atom*
Tetrahedral	3,7-*exo*-tet favoured	5,6-*endo*-tet disfavoured
Trigonal	3,7-*exo*-trig favoured	3,5-*endo*-trig disfavoured
		6,7-*endo*-trig favoured
Digonal	3,4-*exo*-dig disfavoured	5,7-*exo*-dig favoured
		3,7-*endo*-dig favoured

3.3.5 Enol Derivatives

The synthetic carbanion reactions of enolates described above involve firstly the formation of the anion and then the formation of the new carbon–carbon bond. There are some advantages to be gained in terms of regiospecificity and stereospecificity by separating these steps and forming an enol derivative first. If the starting ketone is unsymmetrical, the enol may be formed in a **regiospecific** manner. Furthermore, an open-chain enol can exist as both "*E*" and "*Z*" isomers. Each of these may react in a different manner with an electron-deficient carbon. Consequently, methods have been developed for the regiospecific preparation of particular enolates and for trapping them as enol ethers or enamines.

The conditions for the formation of silyl enol ethers have been adapted for the isolation of either the kinetic or the thermodynamic enolate (see Scheme 3.6). The use of silyl enol ethers in synthesis is illustrated by the sequence in Scheme 3.22.

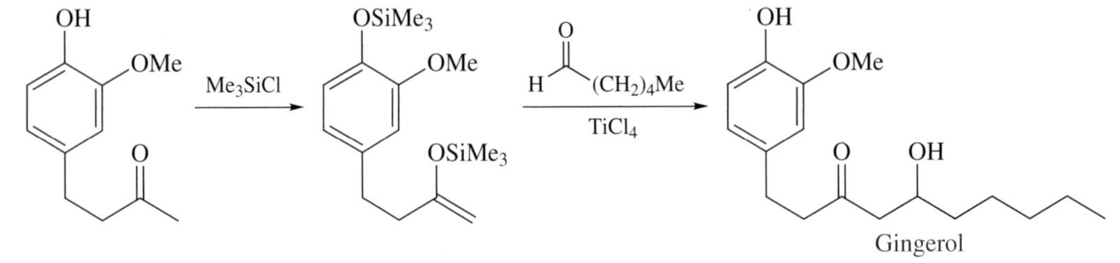

OH, OMe, O; Me_3SiCl → $OSiMe_3$, OMe, $OSiMe_3$; H(C=O)$(CH_2)_4Me$, $TiCl_4$ → OH, OMe, O, OH

Gingerol

Scheme 3.22

3.4 The Stereochemistry of Condensation Reactions

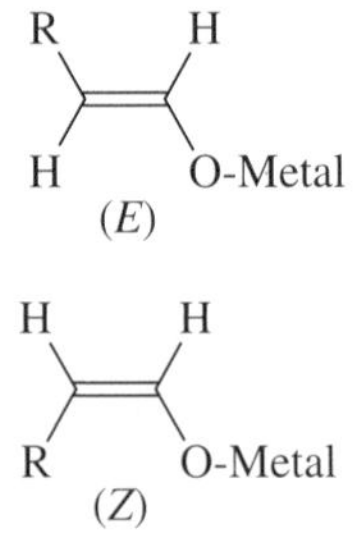

The reaction of a metal salt of an enolate with an aldehyde involves an ordered six-membered chair transition state in which the metal is complexed to both the enolate anion and the oxygen of the aldehyde. The most favoured transition state is that in which the alkyl substituents, R^1–R^3, occupy the equatorial positions. An enolate may be formed with either the "*E*" or the "*Z*" geometry. The "*E*" isomer adds to the aldehyde to give the most stable chair intermediate, which then forms the *anti* diastereomer (Scheme 3.23). On the other hand, the "*Z*" isomer adds to the carbonyl group to form the *syn* diastereomer.

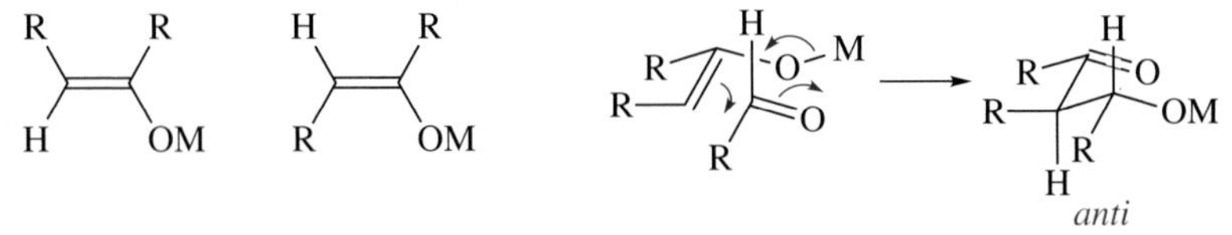

R, H, H, O-Metal (*E*)

H, H, R, O-Metal (*Z*)

R, R, H, OM; H, R, R, OM; H, O, M, R, R, O, R → H, R, O, R, OM, R, H

anti

Scheme 3.23

Zinc enolates can be prepared from bromo esters, as in the **Reformatsky reaction** between a bromo ester and a ketone. In the transition state for these reactions the zinc can coordinate to both oxygens, giving an ordered system (Scheme 3.24).

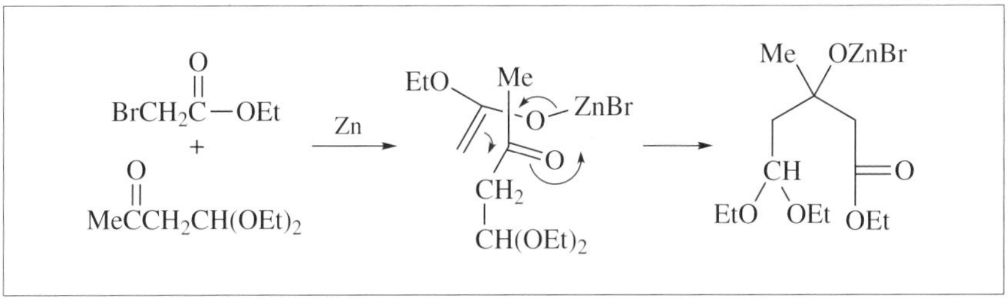

Scheme 3.24 Reformatsky reaction

When the condensation reaction is carried out under equilibrating conditions in which thermodynamic control exists (mild base, long reaction time), the *anti* product predominates. However, under kinetic control (strong base, low temperature, short reaction time), the *syn* diastereomer may be preferred. Since the enolate geometry plays an important role in determining the diastereoselectivity, methods have been developed for trapping and purifying specific enolates as their silyl enol ethers. The enolate anion itself may then be regenerated with fluoride ion under conditions in which it cannot equilibrate with its geometric isomer.

Enol borinates can be prepared by treating a ketone with a dialkylboron trifluoromethanesulfonate in the presence of a base such as *N*,*N*-diethyl-*N*-isopropylamine. Since the boron has a vacant orbital, it can accept the lone pair of electrons from a carbonyl oxygen. This leads to the formation of a highly ordered transition state for an aldol condensation under very mild conditions.

3.4.1 Chiral Enolates

When an enolate is alkylated, although a potentially chiral centre may be generated, normally a racemic mixture will be formed. However, by incorporating a chiral auxiliary group into the enolate, diastereoselective alkylation can be achieved. Once the new chiral centre has been formed, the auxiliary directing group can be removed. Auxiliary groups for this purpose have been obtained from the amino acid valine and the amino alcohol norephedrine (Scheme 3.25).

3.4.2 Enamines

Pyrrolidine enamines are formed from ketones under conditions of thermodynamic control. The tendency is to form the least-substituted enamine, in order to minimize steric interactions and maximize the planarity

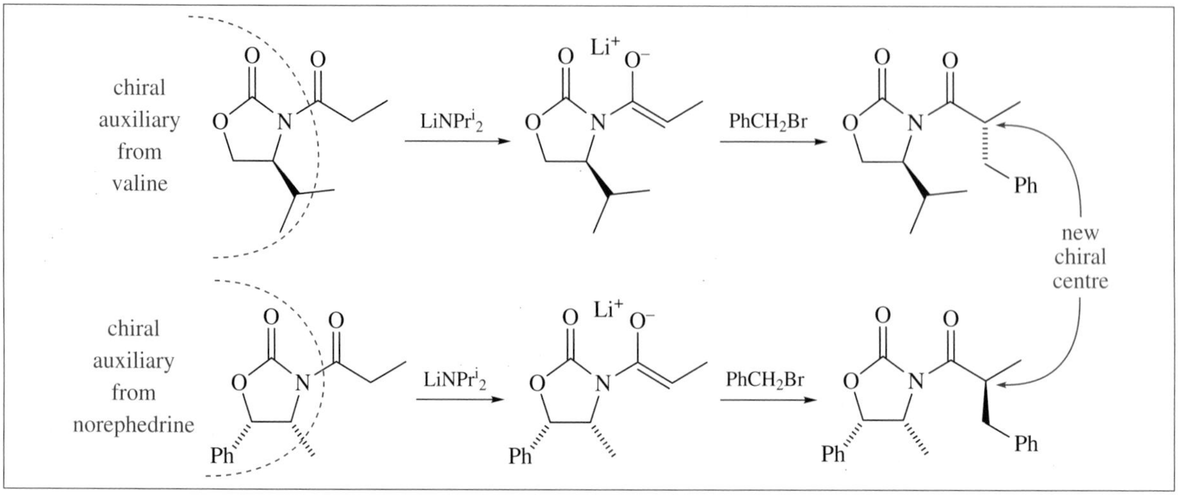

chiral auxiliary from valine

$LiNPr^i_2$

$PhCH_2Br$

new chiral centre

chiral auxiliary from norephedrine

$LiNPr^i_2$

$PhCH_2Br$

Scheme 3.25

of the N–C=C system and hence the overlap of the nitrogen lone pairs with the alkene.

Enamines are alkylated on the β-carbon with alkyl halides under mild conditions to form an imine, which can be hydrolysed with mild acid to the ketone (Scheme 3.26). Enamines condense with aldehydes and ketones and can be acylated with acyl chlorides. The reaction with unsaturated esters has been used in a number of cyclization reactions.

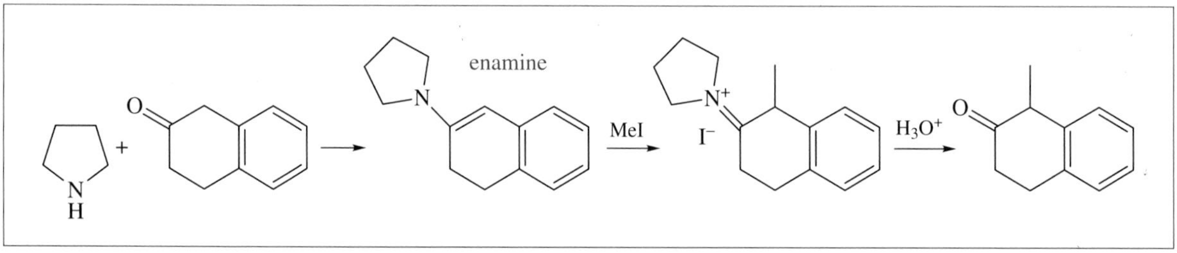

enamine

MeI

H_3O^+

Scheme 3.26

Summary of Key Points

1. A carbonyl group makes an adjacent C–H more acidic. In the presence of base, a carbanion is formed which obtains resonance stabilization by delocalization of the charge through enolate anions.

2. These carbanions behave as nucleophiles, forming C–C bonds in substitution and addition reactions.

3. The major substitution reaction is that of alkylation.

4. The major addition reactions to carbonyl compounds involve aldol condensations, leading to β-hydroxy ketones, Claisen condensations leading to 1,3-diketones and Michael additions leading to 1,5-diketones. The presence of these functional groups in a target molecule is a good "marker" for the use of these reactions in synthesis.

5. Other functional groups such as nitro and nitrile can activate a C–H to carbanion formation.

6. The formation of the enol may be separated from the formation and reaction of enolate anions by the preparation of reactive silyl enol ethers or enamines.

7. The transition states for carbonyl addition reactions can be highly ordered, affording stereochemical control over the products.

Problems

3.1. Outline synthetic routes to the following compounds:

(a) $Me_2C(NO_2)CH_2CH_2CHO$

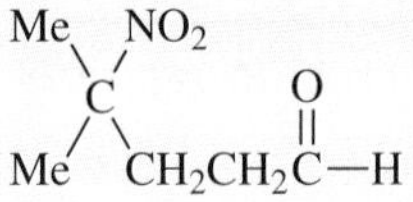

(b) 2,2-dimethyl-3-(CO_2Et)-cyclohexane-1,5-dione (structure)

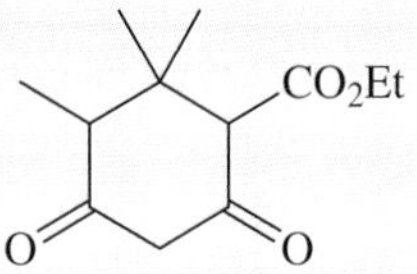

(c) $Me_2C(OH)CH(Me)CO_2Et$

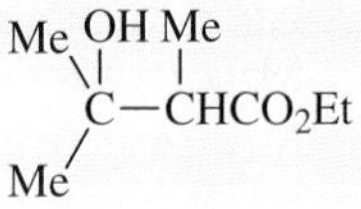

(d) $(EtO_2C)_2CHCH(CO_2Et)CH_2CO_2Et$

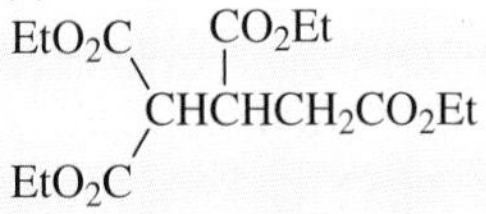

(e) $MeCOCH(Me)CH_2CH{=}CH_2$

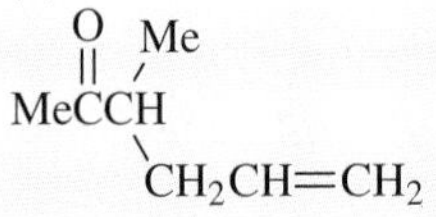

(f) cyclobutane-CO_2H (structure)

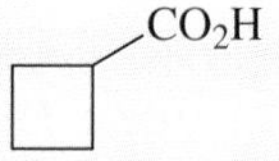

(g) $PhCH_2CH(Me)CO_2H$

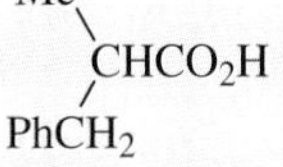

(h) 2-(CO_2Et)-cyclopentanone (structure)

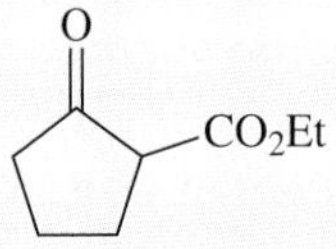

(i) $MeCOCH(I)COMe$

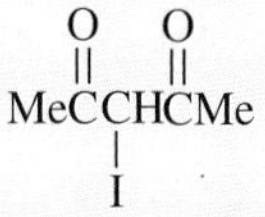

(j) $PhCH{=}CHCOMe$

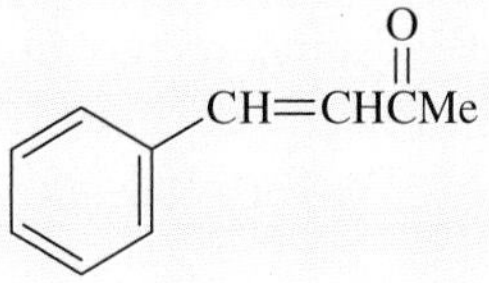

3.2. Using $^{14}CO_2$ as the source of the label, suggest syntheses of the following compounds:

(a) $HO_2CC(OH)HCH_2{}^{14}CO_2H$

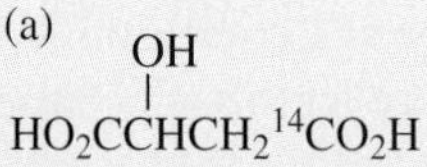

(b) $HO_2{}^{14}C{-}CH(Me){-}CH(Me){-}CH_2{-}CO_2H$

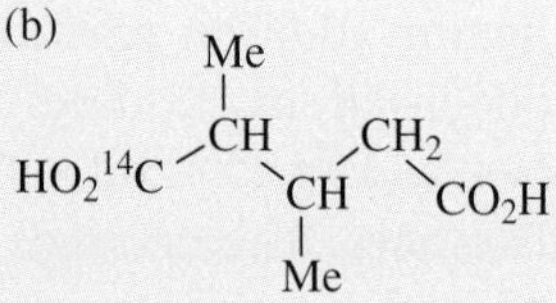

3.3. Suggest syntheses of the following compounds using either diethyl malonate or ethyl acetoacetate as one component:

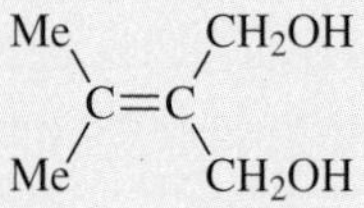

(a) $Me_2C(CH_2OH)_2$

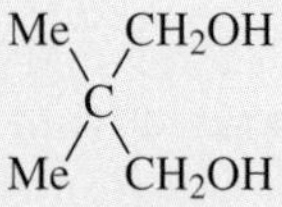

(b) $Me_2C{=}C(CH_2OH)_2$

(c) $MeC(=O){-}CH(Et){-}Et$

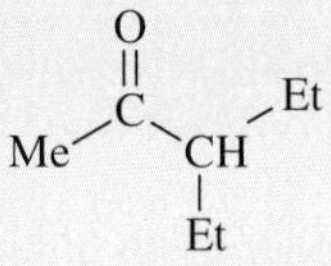

(d) 3-methyl-4-(ethoxycarbonyl)cyclohex-2-en-1-one (O, CO_2Et)

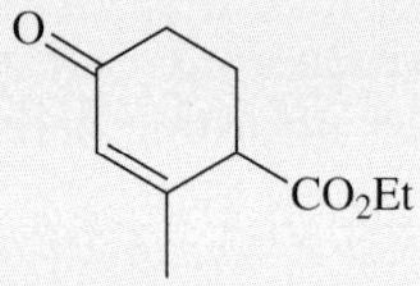

Further Reading

A. T. Nielsen and W. J. Houlihan, *The aldol condensation*, in *Org. React.*, 1968, **16,** 1.

T. Mukaiyama, *The directed aldol reaction*, in *Org. React.*, 1982, **28**, 203.

C. R. Hauser and B. E. Hudson, *The acetoacetic ester condensation*, in *Org. React.*, 1942, **1**, 266.

J. R. Johnson, *The Perkin and related reactions*, in *Org. React.*, 1942, **1**, 210.

G. Jones, *The Knoevenagel condensation*, in *Org. React.*, 1967, **15**, 204.

J. P. Schaefer and J. L. Bloomfield, *The Dieckmann condensation*, in *Org. React.*, 1967, **15**, 1.

W. Johnson and G. H. Daub, *The Stobbe condensation*, in *Org. React.*, 1951, **6**, 1.

M. W. Rathke, *The Reformatski reaction*, in *Org. React.*, 1975, **22**, 423.

M. S. Newman and B. J. Magerlein, *The Darzens glycidic ester condensation*, in *Org. React.*, 1949, **5**, 413.

E. D. Bergmann, D. Ginsburg and R. Pappo, *The Michael reaction*, in *Org. React.*, 1959, **10**, 179.

R. E. Gawley, *The Robinson annelation and related reactions*, in *Synthesis*, 1976, 777.

J. E. Baldwin, *Rules for ring closure*, in *J. Chem. Soc., Perkin Trans. 1*, 1976, 734

J. E. Baldwin, R. C. Thomas, L. J. Kruse and L. Silberman, *Rules for ring closure*, in *J. Org. Chem.*, 1977, **42**, 3846.

J. E. Baldwin and M. J. Lusch, *Rules for ring closure*, in *Tetrahedron*, 1982, **38**, 2939.

J. K. Rasmussen, *O-Silylated enolates, versatile intermediates in synthesis*, in *Synthesis*, 1977, 91.

P. Brownbridge, *Silyl enol ethers in synthesis*, in *Synthesis*, 1983, 1 and 85.

T. Mukaiyama and S. Kobayashi, *Tin enolates in the aldol, Michael and related reactions*, in *Org. React.*, 1994, **46**, 1.

C. J. Cowden and I. Paterson, *Asymmetric aldol reactions using boron enolates*, in *Org. React.*, 1997, **51**, 1.

J. K. Whitesell and M. A. Whitesell, *Alkylation of ketones and aldehydes via their nitrogen derivatives*, in *Synthesis*, 1983, 517.

4
Carbocations in Synthesis

Aims

The aim of this chapter is to describe the various ways of generating a carbocation and of using it in synthesis. By the end of the chapter you should understand:

- The classification of carbocations in terms of their source and the factors that affect their stability
- The characteristic structural features that arise from the use of carbocations in synthesis

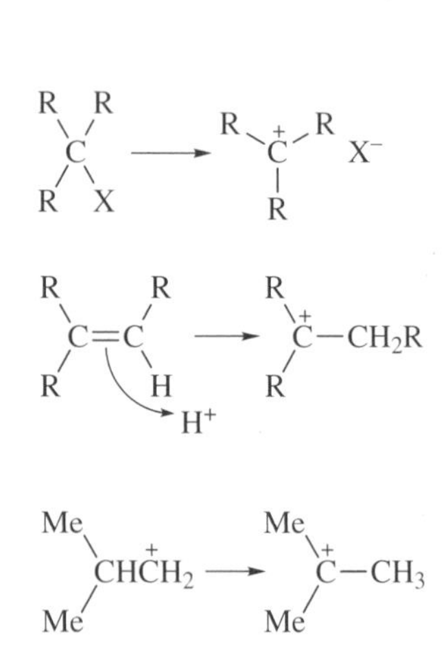

4.1 Introduction

Carbocations react with electron-rich systems such as alkenes, enol ethers, enamines and aromatic rings with the formation of carbon–carbon bonds. Synthetically useful carbocations may be divided into three classes, depending on their oxidation level. Firstly, there are those that are alkyl carbocations. These are obtained by the heterolytic fission of an alkyl halide, alcohol, epoxide or amine or by protonation of an alkene.

If the subsequent reaction of the carbocation is relatively slow, isomerization may take place to generate the more stable tertiary carbocation from a primary or secondary carbocation.

The second family is those that are derived from aldehydes or ketones or by the cleavage of acetals. These carbocations achieve some stabilization from the presence of lone pairs on the oxygen atom. Consequently, they are less likely to undergo rearrangement reactions.

The final group is the acylium ions that are obtained by the cleavage of acid derivatives under Lewis acid catalysis. The resonance stabilization provided by the oxygen lone pairs again restricts the tendency of these acylium ions to rearrange.

There are a number of organometallic compounds involving transition metals in which an electron-deficient centre is generated and with which a carbanion may then react to create a new carbon–carbon bond.

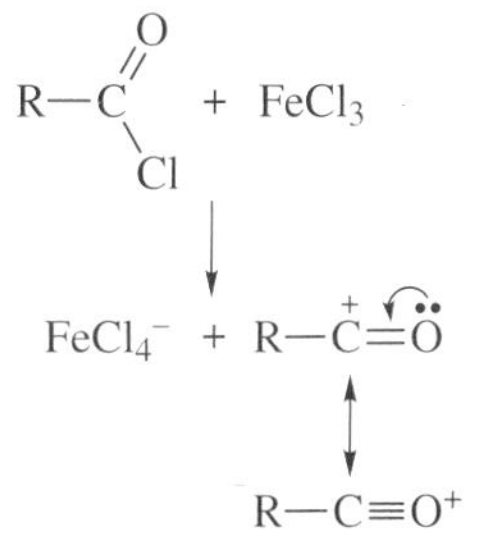

R—C(=O)Cl + $FeCl_3$ → $FeCl_4^-$ + R—$\overset{+}{C}$=Ö ↔ R—C≡O^+

4.2 Alkyl Carbocations: the Friedel–Crafts Alkylation

In the **Friedel-Crafts alkylation** of aromatic rings, the carbocations may be derived from alkenes, alcohols or alkyl halides. The catalysts are Brønsted acids such as sulfuric or phosphoric acids, or Lewis acids such as aluminium trichloride, iron(III) chloride or tin(IV) chloride. The reaction occurs with unsubstituted or electron-rich activated aromatic rings (Scheme 4.1). However, the reaction is unsuccessful with electron-deficient deactivated rings such as nitrobenzene.

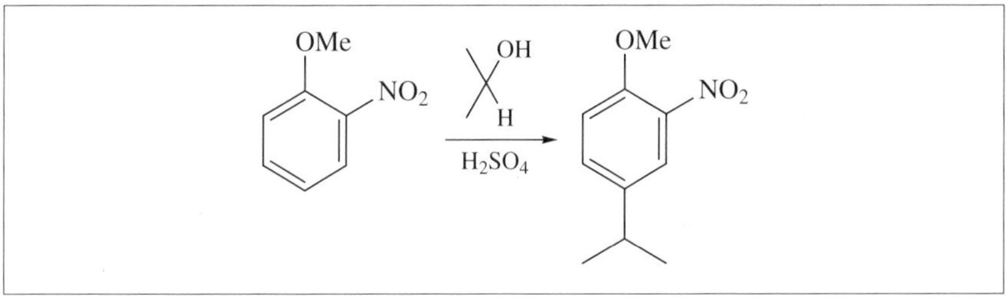

OMe, NO_2, OH, H, H_2SO_4, OMe, NO_2

Scheme 4.1

The ease with which carbonium ion rearrangements take place means that the reaction of benzene with 1-bromopropane gives cumene (isopropylbenzene) rather than propylbenzene. Since an alkylated aromatic ring is more electron rich than the unsubstituted ring, polysubstitution takes place rather easily. The Friedel–Crafts alkylation reactions are reversible, and some isomerization of alkylbenzenes can also take place under Friedel–Crafts catalysis.

The acid-catalysed cyclization of 1,5-dienes provides a method for synthesizing six-membered rings. The cyclization of citral anil to form the cyclocitrals is an example (Scheme 4.2).

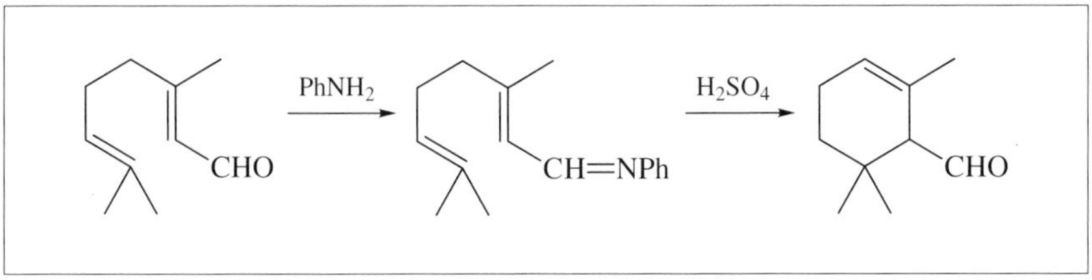

CHO, $PhNH_2$, CH=NPh, H_2SO_4, CHO

Scheme 4.2

4.3 Carbocations Derived from Aldehydes and Ketones

The synthetic use of acetals derived from aldehydes is exemplified by the **Prins reaction** in which a hydroxymethylene group is added to an alkene (Scheme 4.3). The chloromethylation of phenols by formaldehyde (methanal) and hydrochloric acid follows the same pattern. The protonation of the ketonic carbonyl group of β-keto aryl amides and aryl esters provides a synthesis of heterocyclic rings such as quinolines and coumarins (Scheme 4.4).

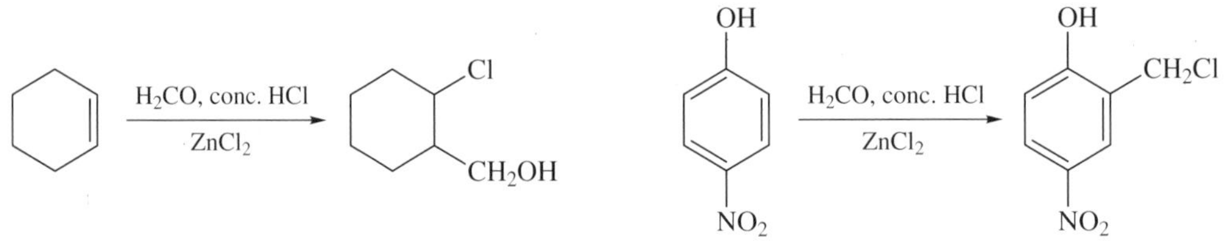

H_2CO, conc. HCl / $ZnCl_2$; Cl, CH_2OH ; OH, NO_2 ; H_2CO, conc. HCl / $ZnCl_2$; OH, CH_2Cl, NO_2

Scheme 4.3

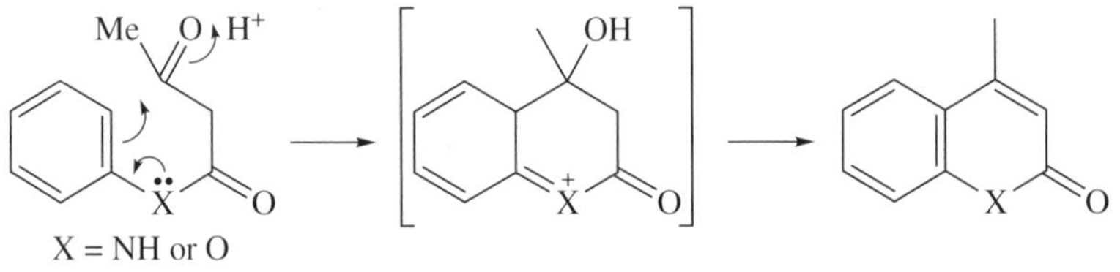

Me, O, H^+ ; OH ; X ; O ; X = NH or O

Scheme 4.4

The **Mannich reaction** between a ketone, formaldehyde and a secondary amine is another example of an acid-catalysed reaction. The reaction involves three components, typically an aldehyde such as formaldehyde, a secondary amine and the enol of a ketone. The acid-catalysed condensation of formaldehyde with a secondary amine such as dimethylamine affords an iminium salt. This behaves as an electrophile, and reacts with the electron-rich enol of a ketone to form a new carbon–carbon bond (Scheme 4.5).

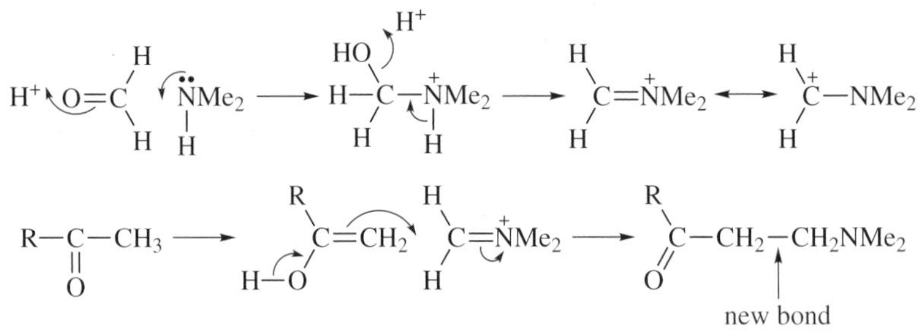

H^+ O=C(H)H + NMe_2H → HO–CH$_2$–$\overset{+}{N}$Me$_2$H (H^+) → $H_2C=\overset{+}{N}Me_2$ ↔ $H_2\overset{+}{C}-NMe_2$

R–C(=O)–CH_3 → R(HO)C=CH_2 + $H_2C=\overset{+}{N}Me_2$ → R–C(=O)–CH_2–CH_2NMe_2 (new bond)

Scheme 4.5 Mannich reaction

The resulting Mannich bases readily undergo elimination reactions to form α,β-unsaturated ketones such as but-3-en-2-one. Iminium salts, such as **Eschenmoser's salt** ($CH_2{=}NMe_2^+ I^-$) may be used in similar reactions with electron-rich alkenes such as enamines. The **Vilsmeier reaction**, which involves the reaction of an alkene with a salt obtained from phosphorus oxychloride and dimethylformamide, follows a similar pattern.

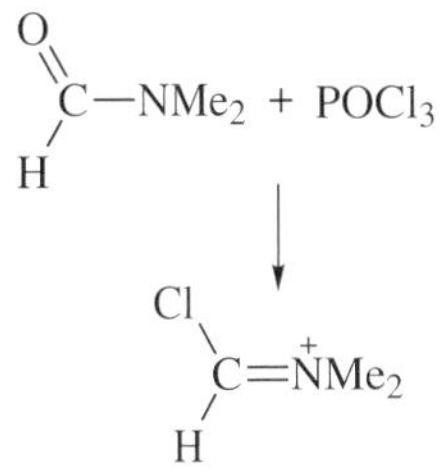

The substitution of aromatic rings by iminium salts is used to form heterocyclic rings. For example, the condensation of a β-phenylethylamine with an aldehyde such as acetaldehyde (ethanal) gives an imine which undergoes acid-catalysed cyclization to form a tetrahydroisoquinoline (Scheme 4.6).

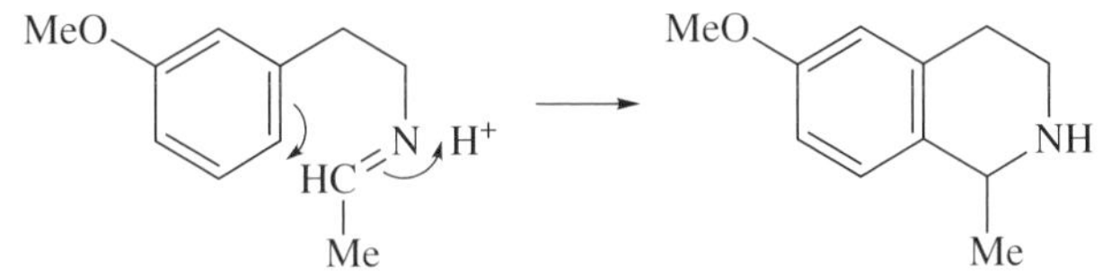

Scheme 4.6

4.4 Acylium Carbocations: the Friedel–Crafts Acylation

An acylium ion can be prepared by treatment of an acyl chloride with a Lewis acid such as iron(III) chloride or aluminium trichloride. Another method involves the reaction of an acid anhydride with concentrated sulfuric acid or polyphosphoric acid. The acylium ion will then react with an alkene or an aromatic ring. Since the acylium ion is less prone to rearrangement, it is possible to attach an alkyl chain to an aromatic ring by acylation, and then to reduce the carbonyl group to a methylene group. The synthesis of the anti-inflammatory drug Ibuprofen (Nurofen) illustrates this (Scheme 4.7).

Scheme 4.7

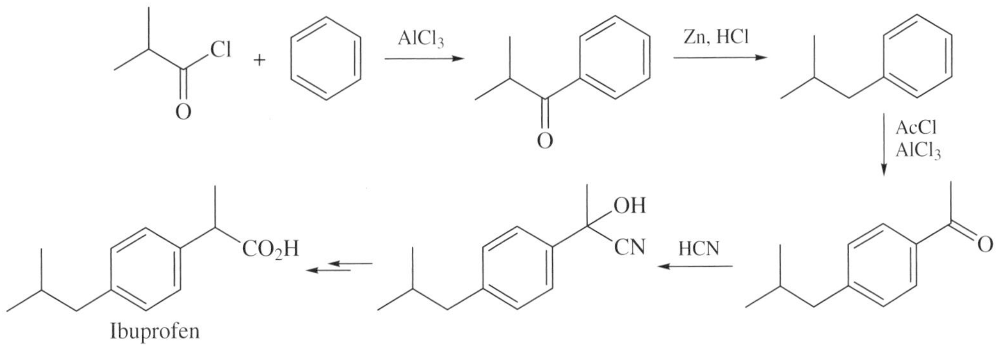

Ibuprofen

A general route for the preparation of various alkylnaphthalenes illustrates the application of acylation reactions in conjunction with other methods of forming carbon–carbon bonds, such as Grignard and alkylation reactions (Scheme 4.8).

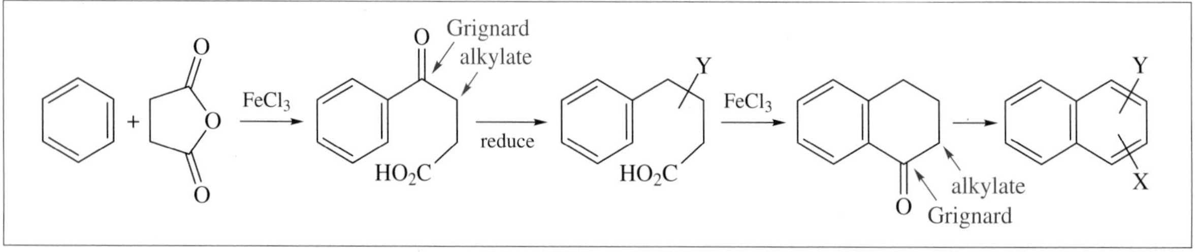

Scheme 4.8

The iminium equivalent of an acylium ion can be formed from acetonitrile and zinc chloride. This has been used in the **Gatterman–Hoesch synthesis** of the flavonoid colouring matters of plants (Scheme 4.9). The electrophile is used to attack the electron-rich aromatic ring of a polyhydroxylic phenol. This reaction is restricted to activated aromatic rings.

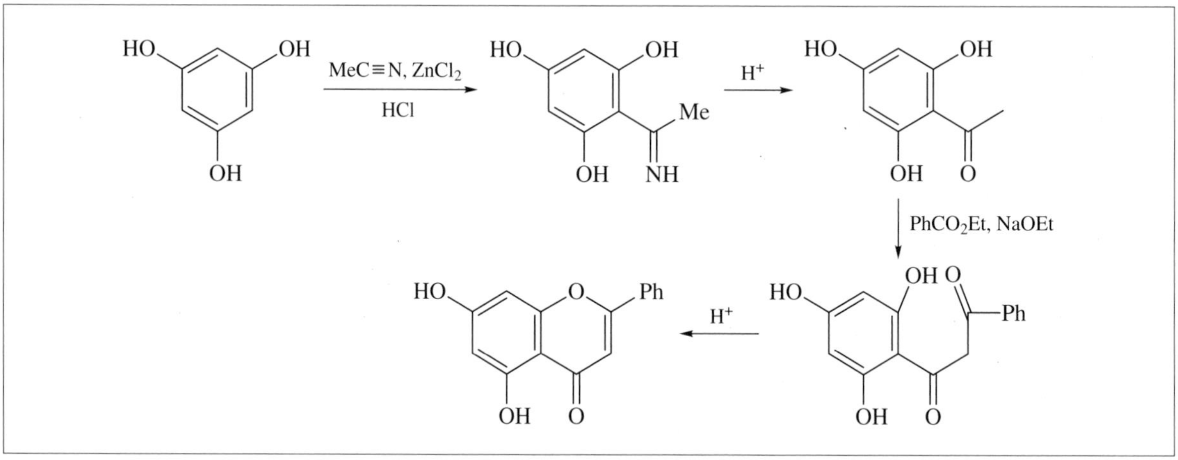

Scheme 4.9
Gatterman–Hoesch reaction

4.5 Acid-catalysed Rearrangement Reactions

A number of rearrangements which involve carbocationic intermediates have some use in synthesis. 1,2-Diols, particularly symmetrical diols (pinacols) which can be prepared by the reductive dimerization of ketones (see Chapter 7), undergo the acid-catalysed **pinacol → pinacolone shift**. This can be used to construct spirocyclic ring systems (Scheme 4.10). In the synthesis of bridged ring systems, it may be easier to prepare one ring system and then make use of **Wagner–Meerwein rearrangements** to convert it to another.

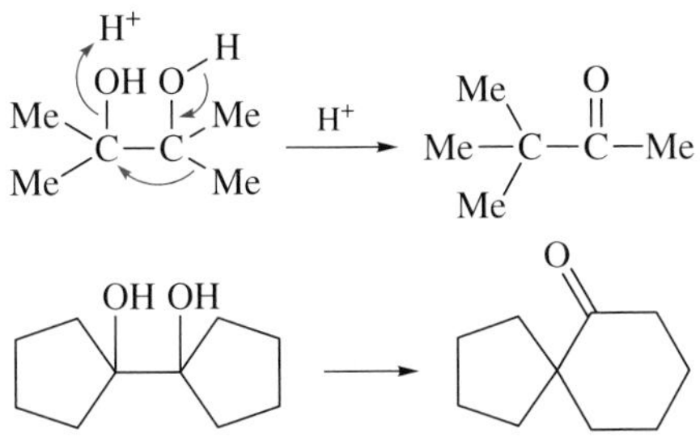

Scheme 4.10 Pinacol–pinacolone rearrangement

Worked Problem 4.1

Q The three 2-, 3- and 4-butylphenols (**1**–**3**) were required during a study of the disinfectant properties of phenols. Suggest how they might be made.

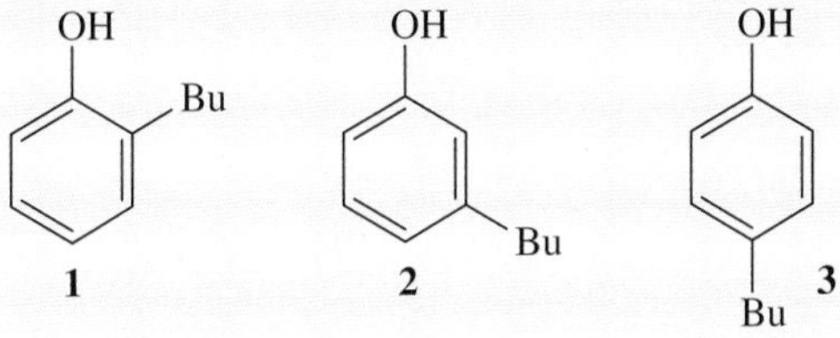

A The straight-chain butyl isomers were required, and hence Friedel–Crafts alkylations of an aromatic ring involving the formation of the primary *n*-butyl carbocation were not suitable because isomerization may occur. Secondly, there is a greater electron density on the oxygen of the phenol than on the aromatic ring, and so to avoid *O*-alkylation the butyl group must be attached to the aromatic ring prior to insertion of the phenol hydroxyl group. Butylbenzene may be prepared either by a Friedel–Crafts acylation of benzene with butanoyl chloride and iron(III) chloride followed by a Clemmensen reduction or by a Fittig reaction using bromobenzene, 1-bromobutane and sodium. Nitration of the butylbenzene gave the *ortho* and *para* isomers which were separated and then converted to the corresponding phenols by reduction to the amine, diazotization and heating the diazonium compound with water. The 3-butylphenol was prepared by nitrating *N*-acetyl-4-butylaniline to give 2-nitro-4-butylacetanilide. The acetanilide was hydrolysed and the free amino group was removed by diazotization and hydrogenolysis of the diazonium sulfate to give 1-butyl-3-nitrobenzene. The nitro group was then converted to the phenol as above.

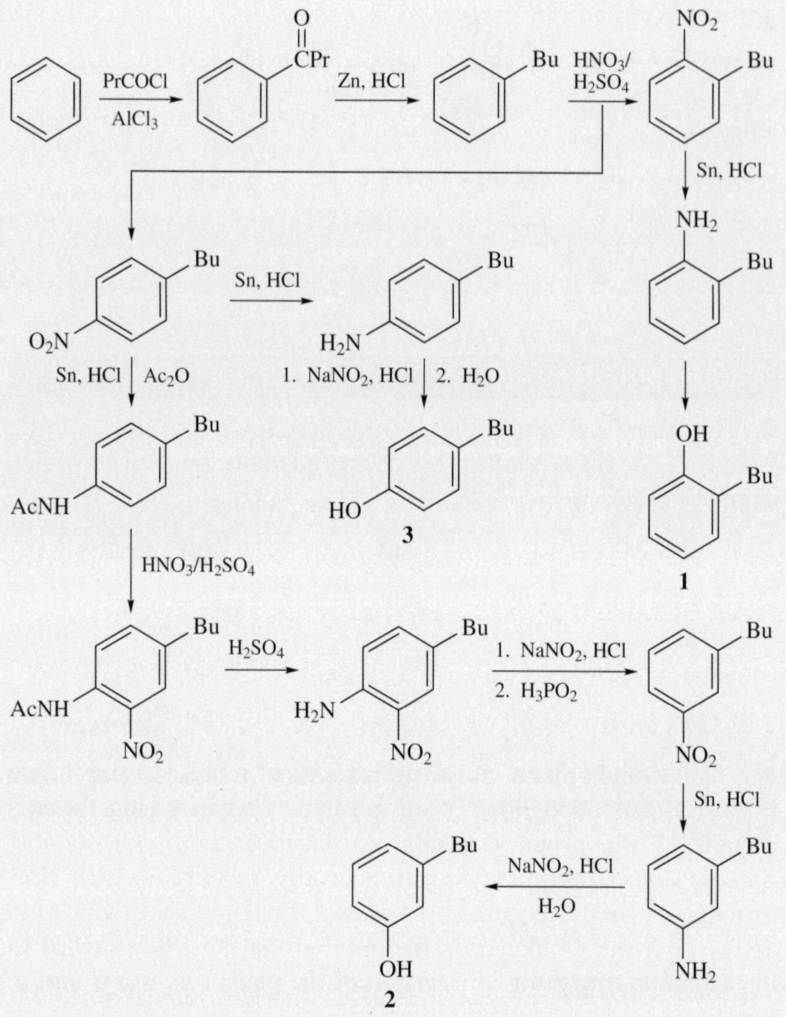

O

CPr

Bu

HNO_3/

H_2SO_4

NO_2

Bu

PrCOCl

$AlCl_3$

Zn, HCl

Sn, HCl

NH_2

Bu

Bu

Sn, HCl

Bu

O_2N

H_2N

Sn, HCl | Ac_2O

1. $NaNO_2$, HCl | 2. H_2O

Bu

Bu

OH

Bu

AcNH

HO

3

1

HNO_3/H_2SO_4

Bu

H_2SO_4

Bu

1. $NaNO_2$, HCl

2. H_3PO_2

Bu

AcNH

H_2N

NO_2

NO_2

NO_2

Sn, HCl

Bu

$NaNO_2$, HCl

H_2O

Bu

OH

2

NH_2

Worked Problem 4.2

Q Propose a synthesis of the (methoxybenzoyl)acrylic acid **4** [(2*E*)-4-(4-methoxyphenyl)-4-oxobut-2-enoic acid].

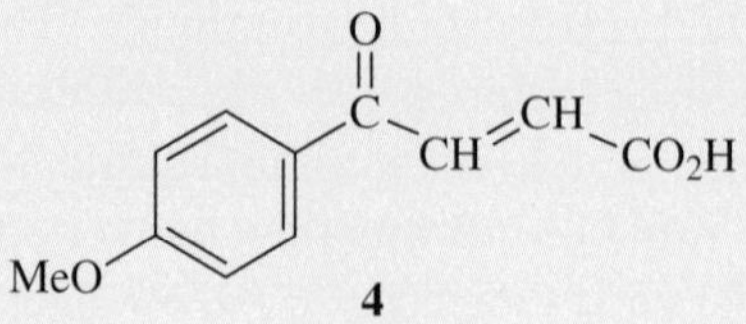

O

C

CH

CH

CO_2H

MeO

4

A The presence of a carbonyl group attached to an aromatic ring, particularly in an activated position *para* to a methoxyl group, suggests a Friedel–Crafts acylation reaction. Maleic anhydride is a suitable acyl unit:

MeO-C6H4 + maleic anhydride —$AlCl_3$→ MeO-C6H4-CO-CH=CH-CO_2H

Worked Problem 4.3

Q Propose a synthesis of 1,3-dihydroxynaphthalene (**5**).

5 (1,3-dihydroxynaphthalene, OH, OH) **6** (1,3-diketone, O, O)

A A phenolic hydroxyl group behaves in some respects as the enol of a ketone. It can be helpful in considering the synthesis of a phenol to write it in a ketonic form. When this is done with 1,3-dihydroxynaphthalene, the 1,3-diketone **6** is revealed. The presence of a carbonyl group adjacent to the benzene ring suggests an acylation reaction. These dissections indicate the following synthetic sequence:

$C_6H_5CH_2COCl$ + $CH_2(CO_2Et)_2$ —NaOEt→ $C_6H_5CH_2C(=O)CH(CO_2Et)_2$ —H_2SO_4→ 2-ethoxycarbonyl-1,3-dioxotetralin (CO_2Et) —1. NaOH, 2. HCl→ **5** (OH, OH)

Summary of Key Points

1. Carbocations may be classified in terms of the oxidation level of their sources.

2. Primary and secondary alkyl carbocations can isomerize to tertiary carbocations, restricting their application in synthesis.

3. Carbocations bearing an oxygen atom and derived from acetals and acyl derivatives are stabilized by the lone pairs on the oxygen and do not rearrange.

4. Carbocations react readily with electron-rich systems such as alkenes and aromatic rings. The Friedel–Crafts alkylation and acylation reactions are major methods of making aromatic C–C bonds. Aryl–carbon bonds are good "markers" for the use of carbocations in synthesis.

Problems

4.1. Suggest syntheses of the following compounds using acid-catalysed reactions:

(a) 2,4,6-trihydroxyphenyl propyl ketone (OH, C(=O)Pr, OH, HO substituted benzene) from phloroglucinol (1,3,5-trihydroxybenzene: OH, HO, OH)

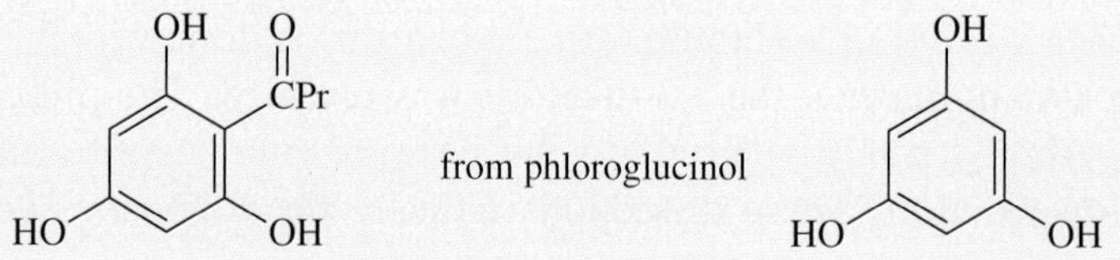

(b) 2-methoxy-5-nitro... benzene bearing OMe, CH_2Cl, NO_2 from anisole (OMe-benzene)

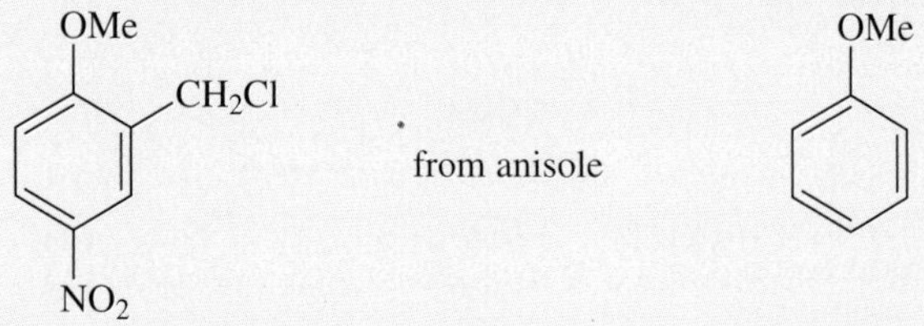

(c) HO-C$_6$H$_4$-C(Me)(Me)CH_2CH_2Me from phenol (OH-benzene)

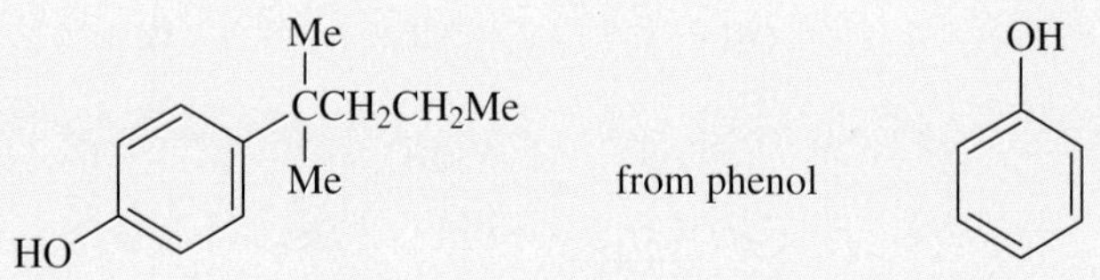

(d) p-$ClC_6H_4CCH{=}CHCO_2H$ (with C=O) from chlorobenzene

(e) Cl–C_6H_4–$C(=CCl_2)$–C_6H_4–Cl from chlorobenzene

4.2. Outline a synthetic route to eudalene (**7**) from benzene.

7

4.3. Show how cyclohexene may be converted into the following compounds:

(a) 1-acetylcyclohexene ($C(=O)Me$) (b) 1-cyclohexenyl-CH_2OH

Further Reading

C. C. Price, *The alkylation of aromatic compounds by the Friedel–Crafts method*, in *Org. React.*, 1946, **3**, 1

J. K. Groves, *The Friedel–Crafts acylation of alkenes*, in *Chem. Soc. Rev.*, 1972, **1**, 73.

E. Berliner, *The Friedel–Crafts acylation of alkenes*, in *Org. React.*, 1949, **5**, 229.

G. Baddeley, *Modern aspects of the Friedel–Crafts reaction*, in *Q. Rev. Chem. Soc.*, 1954, **8**, 355.

F. F. Blicke, *The Mannich reaction*, in *Org. React.*, 1942, **1**, 303.

D. R. Adams and S. P. Bhatnager, *The Prins reaction*, in *Synthesis*, 1976, 661.

A. P. Krapcho, *Synthesis of carbocyclic spiro compounds via rearrangement routes*, in *Synthesis*, 1976, 425.

5
Free Radical and Pericyclic Reactions in the Formation of Carbon–Carbon Bonds

Aims

The aim of this chapter is to illustrate the use of free radical, carbene and pericyclic reactions in the formation of C–C bonds. By the end of this chapter you should understand:

- The steps in the polymerization of alkenes
- The role of the initiator in radical processes
- The role of radicals in coupling processes
- The development of radical cascade sequences
- How to recognize the structural features which arise from carbene and Diels–Alder reactions
- The role of Claisen and Cope rearrangements in the formation of new C–C bonds and recognize the relationship of the new bonds that are formed to double bonds in the target molecules

5.1 Carbon Radical Reactions

A carbon radical is a trivalent species containing an unpaired electron in a p orbital. It is formed by the **homolysis** of a bond between carbon and another atom. Since the homolytic dissociation energies of carbon–carbon, carbon–nitrogen, carbon–oxygen and carbon–halogen bonds are quite high, and there is relatively little external stabilization by solvation, many radical reactions are initiated by the homolysis of weaker bonds and the subsequent transfer of the radical to carbon. Typical **initiators** for this purpose are compounds such as dibenzoyl peroxide and azobisisobutyronitrile (AIBN). The former utilizes the weak oxygen–oxygen bond and the latter the ready formation of nitrogen gas and the stabilization of the resultant radical by delocalization over the nitrile nitrogen (Scheme 5.1).

PhC(=O)–O–O–C(=O)Ph
benzoyl peroxide

$Me_2C(CN)$–N=N–$C(CN)Me_2$
azobisisobutyronitrile (AIBN)

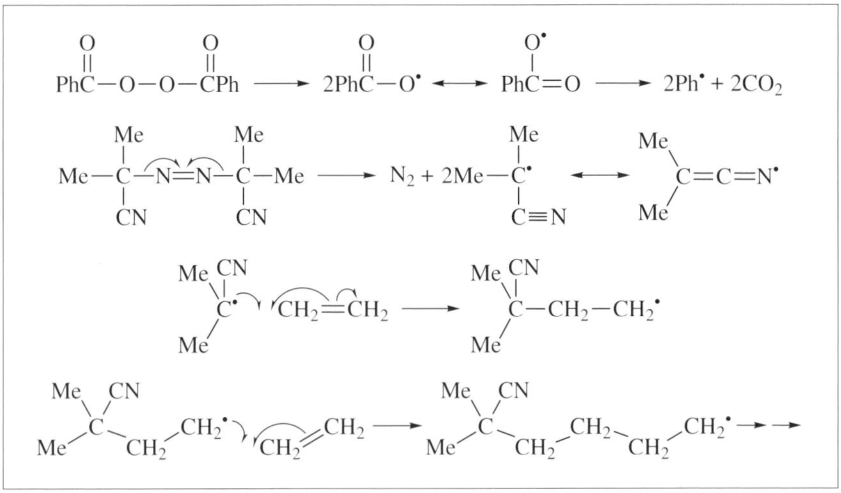

$$PhC(=O)-O-O-C(=O)Ph \longrightarrow 2PhC(=O)-O^{\bullet} \longleftrightarrow PhC(-O^{\bullet})=O \longrightarrow 2Ph^{\bullet} + 2CO_2$$

$$Me-C(Me)(CN)-N=N-C(Me)(CN)-Me \longrightarrow N_2 + 2Me-C^{\bullet}(Me)-C\equiv N \longleftrightarrow Me_2C=C=N^{\bullet}$$

$$Me_2C^{\bullet}(CN) + CH_2=CH_2 \longrightarrow Me_2C(CN)-CH_2-CH_2^{\bullet}$$

$$Me_2C(CN)-CH_2-CH_2^{\bullet} + CH_2=CH_2 \longrightarrow Me_2C(CN)-CH_2-CH_2-CH_2-CH_2^{\bullet} \longrightarrow \longrightarrow$$

Scheme 5.1

The **polymerization** of alkenes illustrates the stages in many radical reactions. The initiator, for example AIBN, dissociates on heating, and the radical adds to an alkene to generate an alkene carbon radical in the initiation step. The new carbon radical then attacks further alkene molecules in the propagation steps. Finally, in the chain termination step the radical may either combine with another developing chain or lose a hydrogen atom.

Synthetically useful radical reactions involve coupling, addition and substitution reactions. One of the oldest coupling reactions is the **Kolbe electrolysis** of carboxylic acids. Discharge of the carboxylate anion by the removal of a single electron leads to the fragmentation of the carboxyl radical, the elimination of carbon dioxide and the formation of alkyl radicals which dimerize. The formation of carbon dioxide facilitated the homolytic fission of the carbon–carbon bond.

The **oxidative coupling of phenols** has attracted considerable theoretical interest. Removal of an electron from a phenolate anion by a one-electron transfer oxidant, such as potassium hexacyanoferrate(III) [$K_3Fe(CN)_6$], generates a radical which may be delocalized over the aromatic ring to give free-radical reactivity at the *ortho* and *para* positions. Coupling of two such radicals can lead to new carbon–carbon bonds (Scheme 5.2).

A number of coupling reactions which involve copper, either as the metal or as a salt, have free-radical characteristics. The **Ullmann coupling** of aryl iodides to give diphenyls and the **Pschorr synthesis** of phenanthrenes follow this pattern (Scheme 5.3). The latter involves the loss of nitrogen from a diazonium salt. A related coupling of diazonium salts also gives rise to biphenyls.

The **Cadiot–Chodkiewicz oxidative coupling** of terminal alkynes using ammoniacal copper(I) chloride provides a useful way of building up carbon chains (Scheme 5.3).

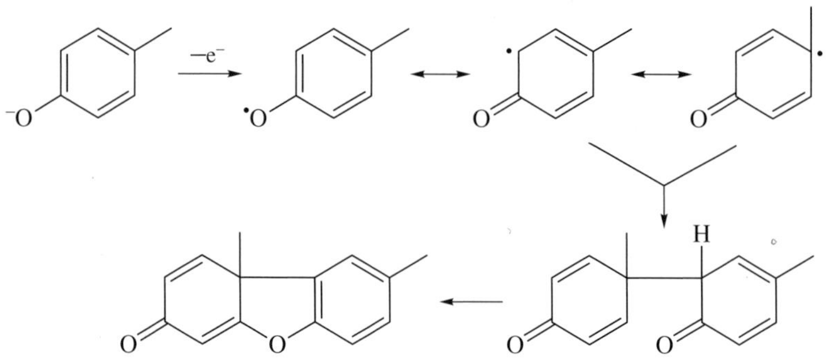

Scheme 5.2

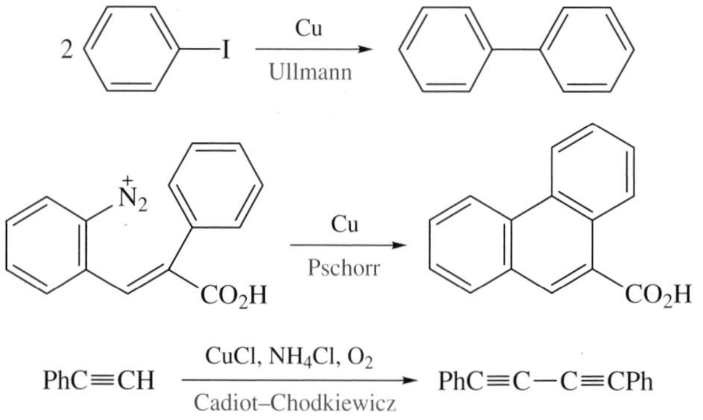

Scheme 5.3 The Ullmann reaction, Pschorr reaction and Cadiot–Chodkiewicz coupling

A number of reductive processes involve the generation of carbon radicals which can couple, with the formation of a new carbon–carbon bond. Magnesium reduces ketones *via* one-electron transfer steps. Since magnesium coordinates strongly to oxygen, it can hold two radicals sufficiently close to facilitate the formation of a dimer, a 1,2-diol or **pinacol** (Scheme 5.4). A good marker for the use of a radical-coupling reaction in a synthesis is the presence of an element of symmetry in the target molecule.

The **acyloin condensation** of esters involving the reduction of the ester with sodium can be used to form ring systems (Scheme 5.4).

The **McMurry coupling** of aldehydes and ketones to form alkenes, using titanium in a low oxidation state, is another example of this principle. This reaction has been particularly successful in making large rings as in a synthesis of humulene (Scheme 5.4).

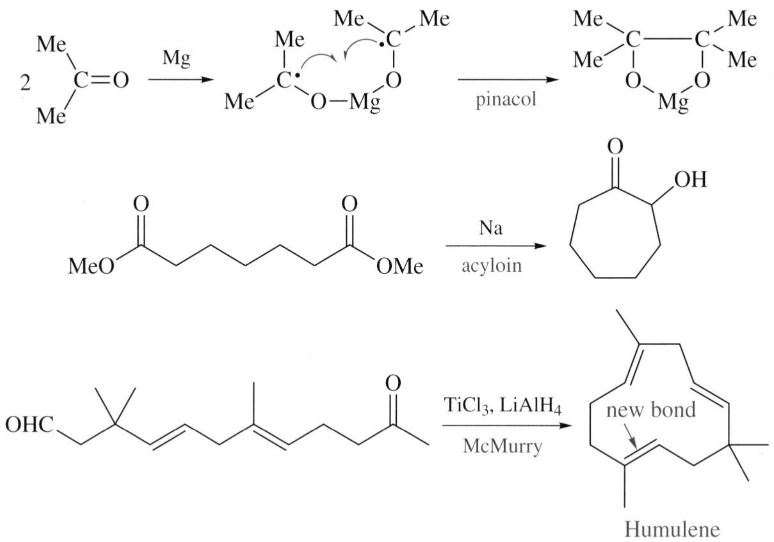

Mg
pinacol
Na
acyloin
$TiCl_3$, $LiAlH_4$
McMurry
new bond
Humulene

Scheme 5.4 Pinacol coupling, acyloin condensation and McMurry coupling

Worked Problem 5.1

Q Devise a synthesis of the *cis*-cyclobutane-1,2-diol (**1**).

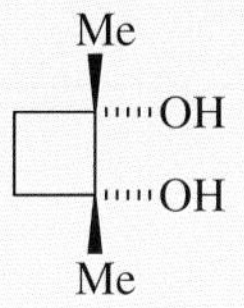

A The target molecule is a symmetrical *cis*-1,2-diol. This would imply a pinacol-type coupling. The synthesis involves the reductive coupling of hexane-2,5-dione with magnesium amalgam and titanium(IV) chloride. The hexane-2,5-dione is obtained from 2,5-dimethylfuran.

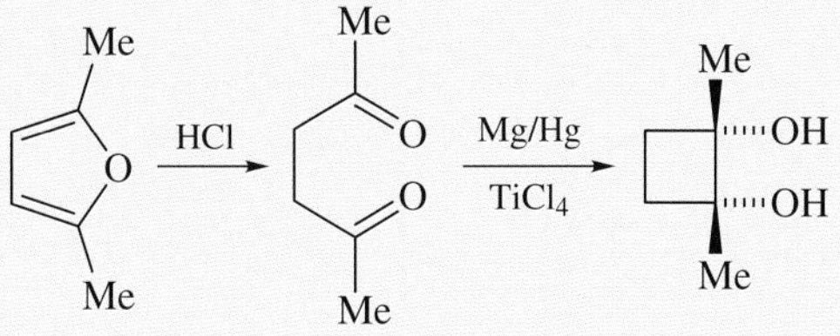

HCl
Mg/Hg
$TiCl_4$

5.2 Radical Addition Reactions

The tin–hydrogen bond of tributylstannane (tributyltin hydride, Bu_3SnH) is relatively weak and will undergo fission in the presence of AIBN. The tin radical may then abstract a halogen atom from a carbon–halogen bond to generate a carbon radical. This carbon radical can then take part in addition reactions with appropriately placed radicophilic centres such as alkenes. The new carbon radical may then remove a hydrogen atom from a further molecule of tributylstannane, so completing the cycle (Scheme 5.5).

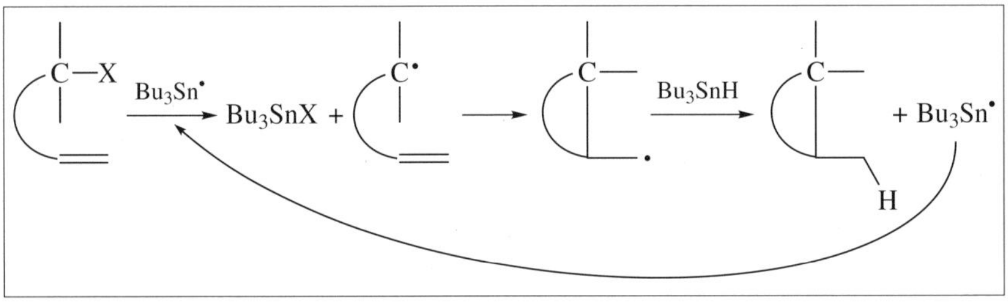

C—X
$Bu_3Sn^\bullet$
Bu_3SnX +
C•
C—
Bu_3SnH
C—
H
+ $Bu_3Sn^\bullet$

Scheme 5.5

These cyclizations often lead to five-membered rings. An important radical in this context is the hex-5-enyl radical. A marker for its use in synthesis can be a five-membered ring with a methyl group as a substituent (Scheme 5.6).

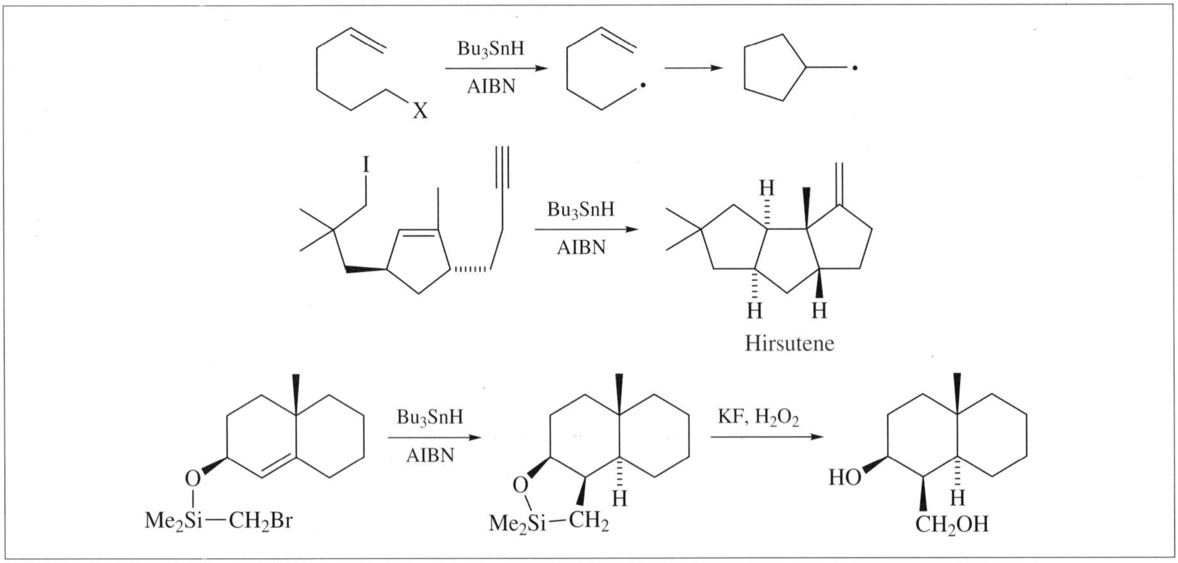

X
Bu_3SnH
AIBN
I
Bu_3SnH
AIBN
H
H H
Hirsutene
O
$Me_2Si—CH_2Br$
Bu_3SnH
AIBN
O
H
$Me_2Si—CH_2$
KF, H_2O_2
HO
H
CH_2OH

Scheme 5.6

Some examples of these additions and cyclizations are given in Scheme 5.6. In some cases a cascade of cyclizations has arisen, as in the synthesis of the sesquiterpenoid hirsutene.

5.3 Carbenes

A **carbene** (R_2C:) is a neutral species in which the carbon atom has only six valence electrons. There are two in each of the bonds, and two non-bonding electrons which can be paired in a singlet carbene or unpaired in a triplet carbene. A carbene is a highly electron-deficient species. In order to achieve the octet, it must participate in the creation of two further bonds. Consequently, many of the reactions of carbenes involve an insertion of the carbene between two atoms or across a π-bond.

Carbenes are formed by a σ-elimination in which two bonds are broken from the same carbon atom. Dichlorocarbene (:CCl_2) is formed from chloroform (trichloromethane) by reaction with a strong base (*e.g.* NaOH) (Scheme 5.7). The elimination of iodine from diiodomethane by a zinc/copper couple gives a zinc-stabilized carbene (Scheme 5.7). **Diazomethane** also loses nitrogen to generate a carbene. The decomposition of a number of aliphatic diazo compounds, particularly diazo ketones, also generates carbenes.

$$\bar{C}H_2{-}\overset{+}{N}{\equiv}N \longrightarrow CH_2 + N{\equiv}N$$

diazomethane

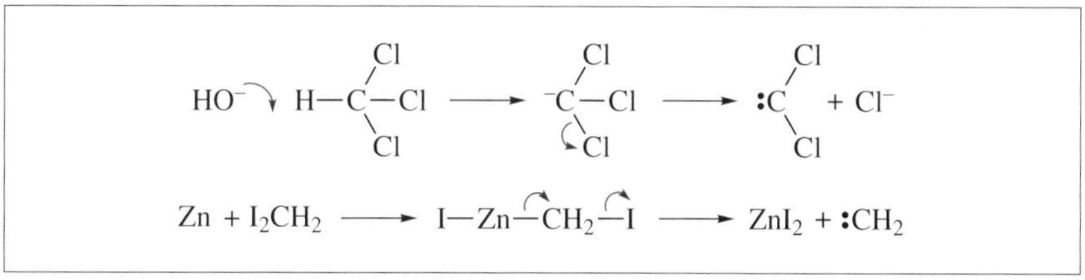

$$HO^- \;\; H{-}CCl_3 \longrightarrow {}^-CCl_3 \longrightarrow :CCl_2 + Cl^-$$

$$Zn + I_2CH_2 \longrightarrow I{-}Zn{-}CH_2{-}I \longrightarrow ZnI_2 + :CH_2$$

Scheme 5.7

The high reactivity of carbenes towards insertion into the π-system of alkenes is reflected by their widespread use in making cyclopropane rings. The **Simmons–Smith methylenation** reaction is an example of this. An application of a diazo ketone in this context is in the synthesis of ethyl chrysanthemate (Scheme 5.8).

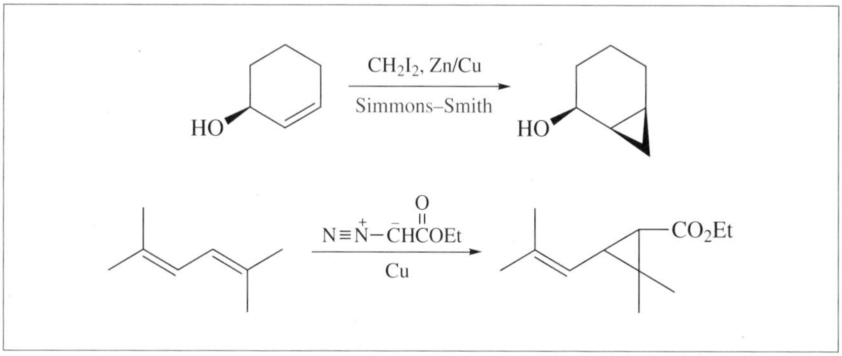

CH_2I_2, Zn/Cu; Simmons–Smith

$N{\equiv}\overset{+}{N}{-}\overset{-}{C}HCOEt$ (O), Cu; CO_2Et

Scheme 5.8 Simmons–Smith methylenation

The chain lengthening of carboxylic acids by the **Arndt–Eistert reaction** is an example of a carbene reaction based on diazomethane

(Scheme 5.9). Diazomethane reacts with an acyl chloride to give a diazo ketone. This decomposes under the influence of a silver salt with the loss of nitrogen to give a carbene which rearranges to a ketene. Hydration of the ketene leads to an acid.

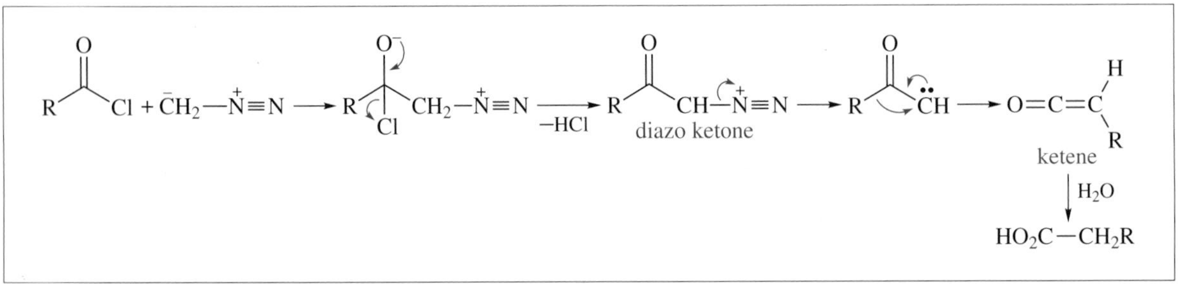

diazo ketone

ketene

H_2O

HO_2C-CH_2R

Scheme 5.9 The Arndt–Eistert reaction

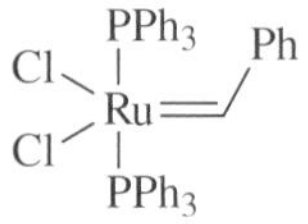

Grubb's catalyst

5.4 Alkene Metathesis

Some metals such as ruthenium will stabilize carbenes as complexes. For example, ruthenium forms a complex with phenylcarbene. This ruthenium complex and some relatives will catalyse a reaction known as alkene metathesis. In these reactions between two alkenes, a=b and c=d, the bonding partners are interchanged to give a=c and b=d. The synthetic value of the reaction is when one of products (*e.g.* ethene) is easily separated from the other product and the reaction can be driven to completion. The reactions involve the formation of four-membered metallocyclobutanes (Scheme 5.10), which comprise interchange of a carbene complex and the removal of one carbon, in this case the terminal carbon of styrene. The new ruthenium carbene complex may then repeat

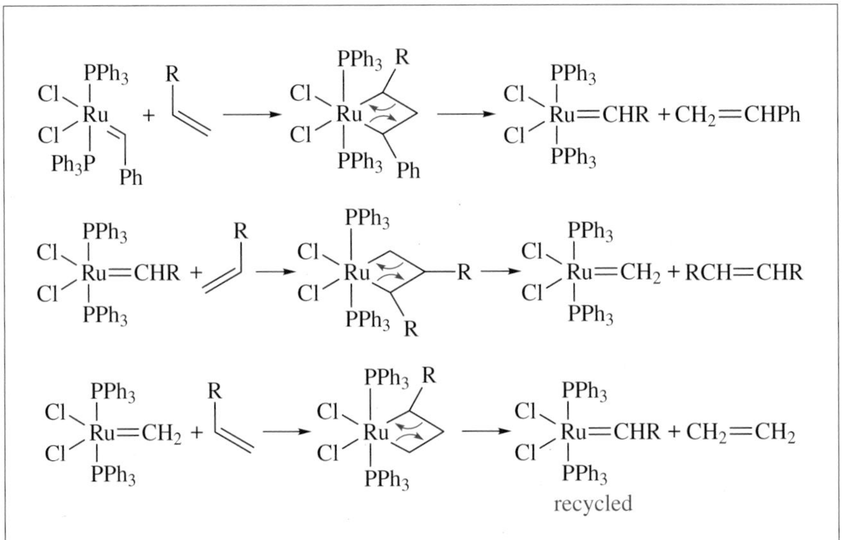

recycled

Scheme 5.10

the cycloaddition (Scheme 5.10). The outcome of this cycle of reactions is the extrusion of ethene from the terminal carbons and the coupling of two chains *via* an alkene. A major use of this reaction is in a cyclization known as **ring-forming metathesis** (Scheme 5.11).

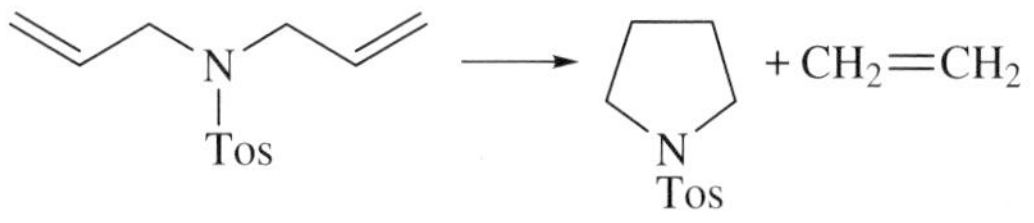

N Tos → N Tos + $CH_2{=}CH_2$

Scheme 5.11 Ring-forming metathesis

5.5 The Diels–Alder Reaction

The **Diels–Alder reaction** (Scheme 5.12) between a **diene** and a **dienophile** is one of the major synthetic methods for the creation of new carbon–carbon bonds. The regio- and stereochemistry of the products are determined by the principles of the conservation of orbital symmetry. The thermal [4 + 2] cycloaddition of a diene with a dienophile is exemplified by the addition of cyclopentadiene to maleic anhydride (*cis*-butenedioic acid anhydride) and to benzo-1,4-quinone (Scheme 5.13). Note the need for the dienophile to contain an electron-withdrawing group. These examples illustrate the value of the process in that two new bonds of defined geometry are created. A Diels–Alder synthesis is particularly useful in creating bridged and *cis*-fused ring junctions and in forming six-membered rings.

A [4 + 2] cycloaddition involves four electrons from one reactant and two electrons from the other.

Scheme 5.12 Diels–Alder cycloaddition

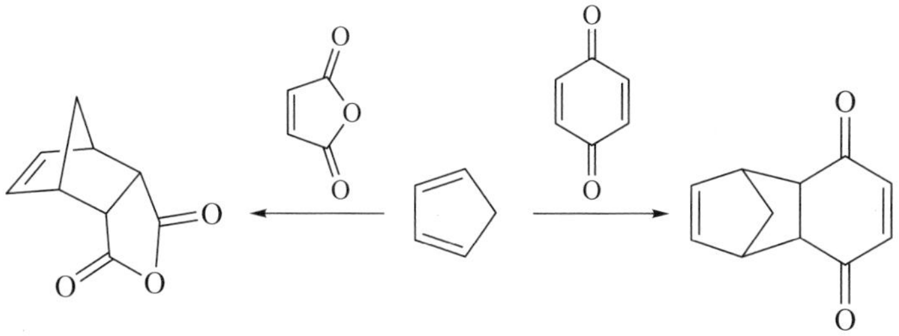

Scheme 5.13

The stereochemistry of a substituent on the cyclopentadiene may direct the face of addition of the dienophile. The lactone in Scheme 5.14 has played an important role in the synthesis of the prostaglandins.

The stereochemistry of the addition of acetoxyacrylonitrile to (methoxymethyl)cyclopentadiene was determined by the methoxymethyl substituent, and thus the stereochemistry of three out of the four asymmetric centres of the lactone was created in one step.

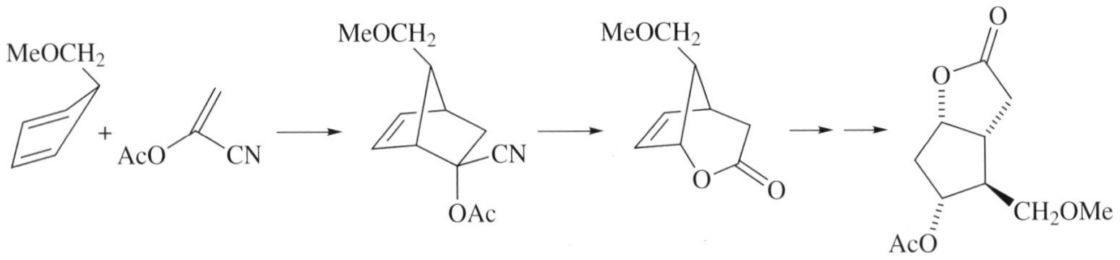

MeOCH$_2$ + AcO CN → MeOCH$_2$ CN OAc → MeOCH$_2$ O O → → O O AcO CH$_2$OMe

Scheme 5.14

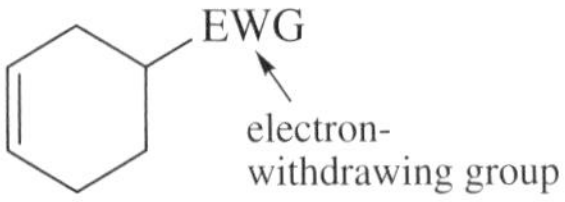

The presence of a cyclohexene and an appropriately placed electron-withdrawing group are **markers** for the use of the Diels–Alder reaction in synthesis.

EWG
electron-
withdrawing group

Worked Problem 5.2

Q Propose a synthesis of the cyclohexene **2**.

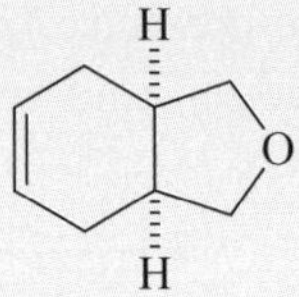

H O **2** H

A The presence of a cyclohexene ring and a *cis* ring junction between the cyclohexene and the tetrahydrofuran are characteristics of a Diels–Alder reaction. A Diels–Alder dissection reveals butadiene as the diene component. However, the dienophile requires an electron-withdrawing substituent such as a carbonyl group. The synthesis uses butadiene and maleic anhydride followed by reduction of the anhydride:

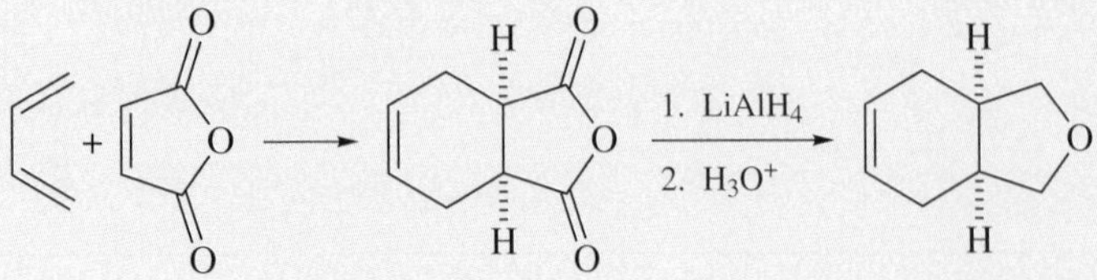

O O O + → H O O O H O → 1. LiAlH$_4$ 2. H$_3$O$^+$ → H O H

5.6 The Ene Reaction

The **ene reaction** is a sigmatropic rearrangement which can lead to the formation of a new carbon–carbon bond (Scheme 5.15). The retention of optical activity in the reaction between an optically active alkene and maleic anhydride indicates the concerted nature of the reaction.

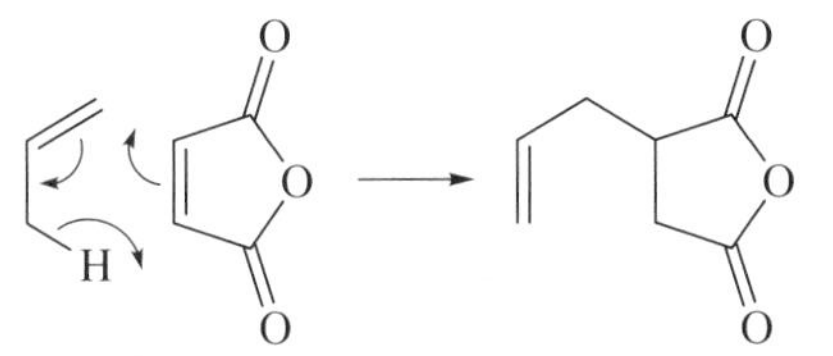

O O O H O O O

Scheme 5.15 The ene reaction

5.7 The Cope and Claisen Rearrangements

The **Cope rearrangement** is a 3,3-sigmatropic reaction of 1,5-dienes which also leads to the creation of a new carbon–carbon bond (Scheme 5.16). A useful variation of this synthetic strategy is the **oxy-Cope rearrangement** of 3-hydroxy-1,5-dienes. The initial product of the rearrangement is an enol, which then gives an aldehyde or ketone (Scheme 5.16).

A sigmatropic rearrangement is a pericyclic reaction in which a σ-bond is broken in the reactant and a new σ-bond is formed in the product whilst a π-bond rearranges.

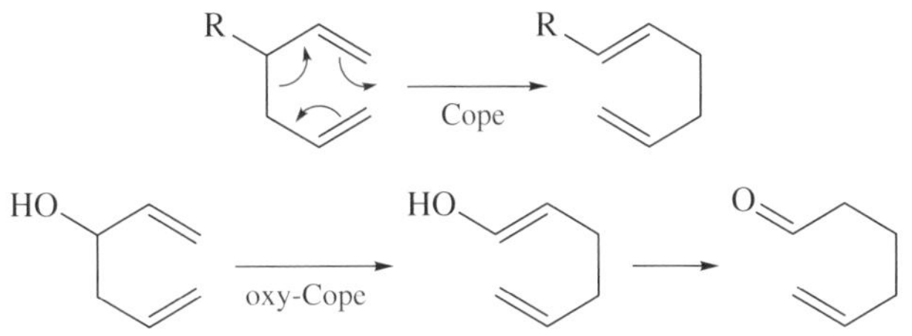

R Cope R
HO oxy-Cope HO O

Scheme 5.16 Cope and oxy-Cope rearrangements

Some typical examples of these rearrangements are found in the reactions of sesquiterpenes and in the synthesis of medium-sized rings (Scheme 5.17).

If an oxygen atom forms part of the 1,5-diene, a carbonyl group may be generated in the rearrangement. The 1,5-diene in the 3,3-sigmatropic rearrangement may also be part of an aromatic system, as in the **Claisen rearrangement** of allyl phenol ethers. Eugenol may be prepared from guiacol allyl ether by this route (Scheme 5.18). It is important to note that the γ-carbon atom of the allyl ether becomes attached to the aromatic ring. In the **Ireland–Claisen rearrangement** a silyl ketene acetal is generated from an allyl ether or ester. The consequence of the 3,3-sigmatropic rearrangement is then an unsaturated acid. The ordered nature

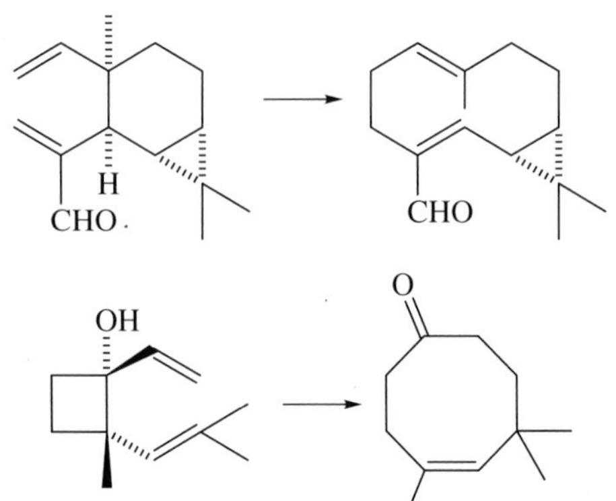

H
CHO
CHO
OH
O

Scheme 5.17

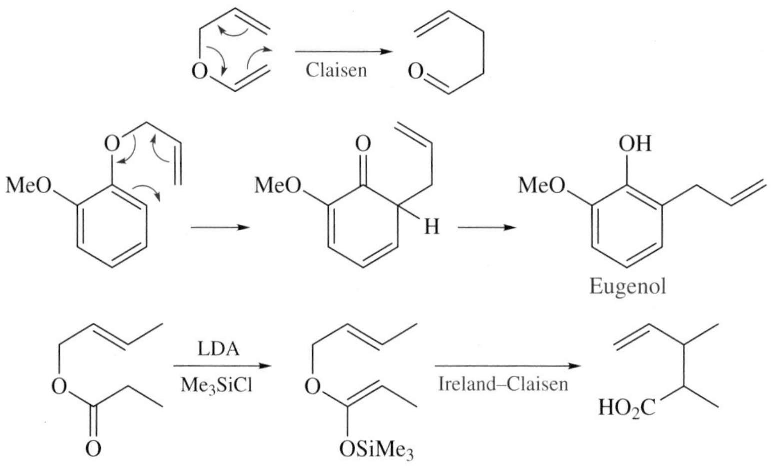

O Claisen O
MeO O MeO O H MeO OH
Eugenol
O LDA Me_3SiCl O Ireland–Claisen HO_2C
O $OSiMe_3$

Scheme 5.18 Claisen and Ireland–Claisen rearrangements

of the transition states in these reactions provides considerable stereochemical control over their outcome.

Summary of Key Points

1. Radical chain reactions involve a number of stages, including initiation, chain propagation and termination.

2. Radical pairing reactions involve the coupling of two radicals and may lead to symmetrical dimeric products.

3. If a substrate contains several radicophilic centres within the molecule, initiation of a radical may lead to a cascade of reactions.

4. Cyclization of the hex-5-enyl radical provides a useful way of making five-membered rings.

5. Carbene reactions lead to cyclopropane rings and to methylene insertion reactions.

6. Alkene metathesis involves the exchange of alkene partners in the sense a=b + c=d → a=c + b=d and can be used to create cyclic alkenes.

7. The Diels–Alder reaction of a diene and a dienophile leads to the formation of cyclohexenes in a stereochemically defined manner.

8. 3,3-Sigmatropic rearrangements, including the Cope, Claisen and Ireland–Claisen reactions, provide useful ways of making new C–C bonds.

Problems

5.1. The following compounds have been synthesized by a Diels–Alder reaction. Identify the diene and dienophile components.

(a) CO_2Me (b) CHO (c) CO_2Me, O, CO_2Me

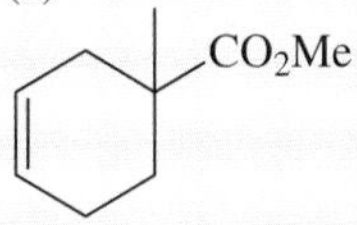

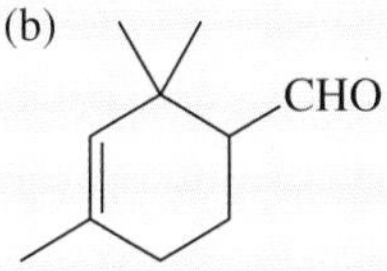

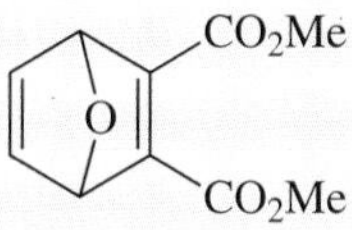

(d) NO_2, Ph (e) O, O (f) CO_2H, Ph

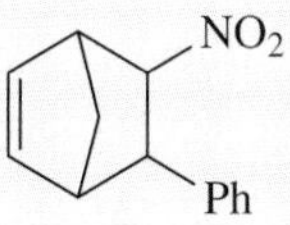

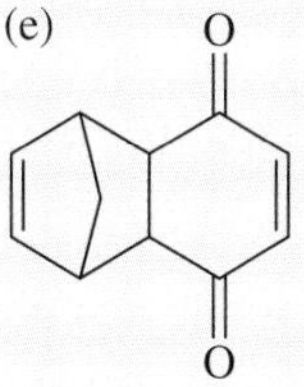

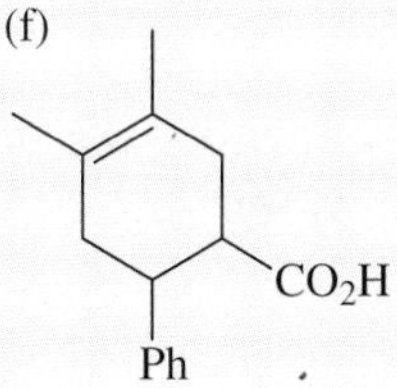

5.2. Show how a carbene might be used in the synthesis of the following compound:

O

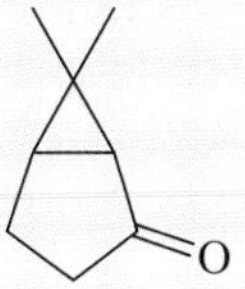

5.3. Give the products of the following ene reactions:

(a)  + $HC{\equiv}CCO_2Me$ $\xrightarrow{200\ °C}$

(b) 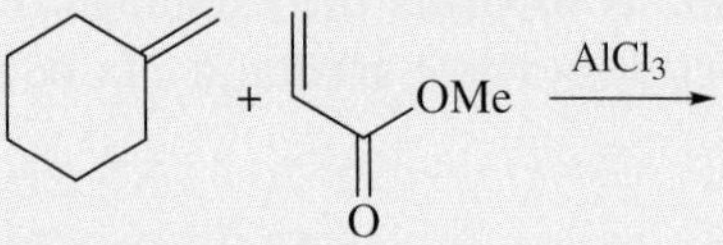 + OMe, O $\xrightarrow{AlCl_3}$

5.4. Give the products of the following radical reactions:

(a) 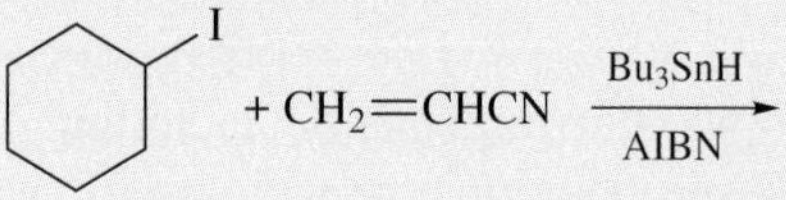 I + $CH_2{=}CHCN$ $\xrightarrow[AIBN]{Bu_3SnH}$

(b) 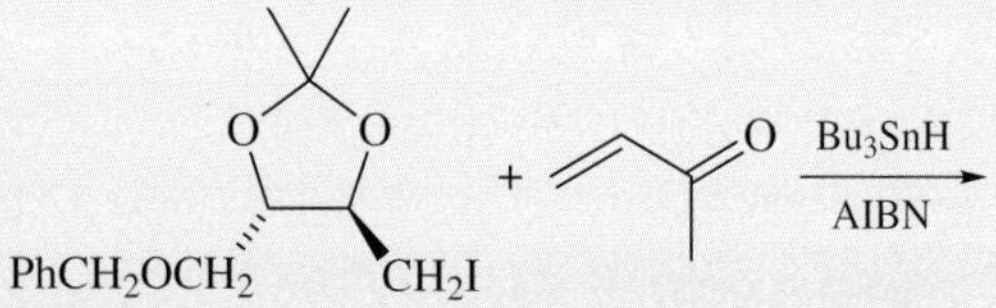 O O, $PhCH_2OCH_2$, CH_2I + O $\xrightarrow[AIBN]{Bu_3SnH}$

(c) 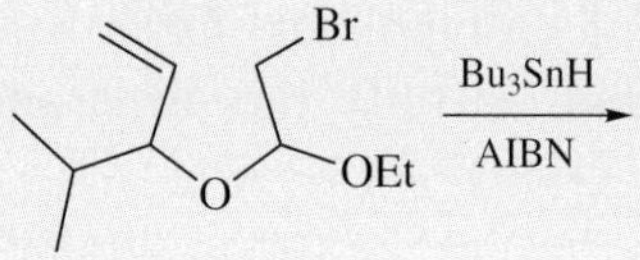 Br, O, OEt $\xrightarrow[AIBN]{Bu_3SnH}$

(d) 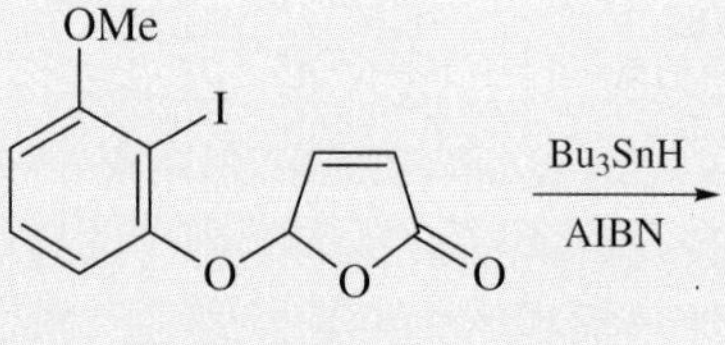 OMe, I, O, O, O $\xrightarrow[AIBN]{Bu_3SnH}$

5.5. Identify the individual steps in the following radical cascade cyclization:

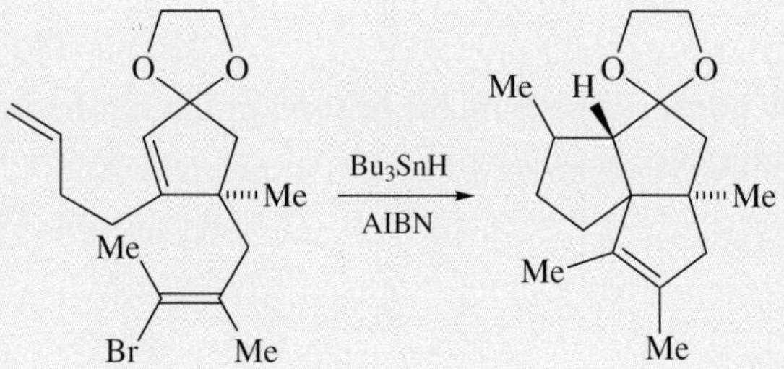

O O, Me, Me, Br, Me $\xrightarrow[AIBN]{Bu_3SnH}$ Me H O O, Me, Me, Me

Further Reading

M. J. Perkins, *Radical Chemistry: the Fundamentals*, Oxford University Press, Oxford, 2000.

D. P. Curran, *The design and application of free radical chain reactions in organic synthesis*, in *Synthesis*, 1988, 417 and 489.

A. L. J. Beckwith, *The pursuit of selectivity in radical reactions*, in *Chem. Soc. Rev.*, 1993, **22,** 143.

G. Majetich and K. Wheless, *Remote intramolecular free radical functionalization, an update*, in *Tetrahedron*, 1995, **51**, 7095.

A. I. Scott, *Oxidative coupling of phenolic compounds*, in *Q. Rev. Chem. Soc.*, 1965, **19**, 1.

J. E. McMurry, *Titanium-induced dicarbonyl-coupling reactions*, in *Acc. Chem. Res.*, 1983, **16**, 405.

D. Lenoir, *Applications of low-valent titanium reagents in organic synthesis*, in *Synthesis*, 1989, 883.

B. Giese, B. Kopping, T. Gobel, J. Dickhaut, G. Thoma, K. J. Kulicke and F. Trach, *Radical cyclization reactions*, in *Org. React.*, 1996, **48**, 301.

S. H. Pine, *Carbonyl methylenation and alkylidenation using titanium based reagents*, in *Org. React.*, 1993, **43**, 1.

C. D. Gutsche, *Reactions of diazomethane and its derivatives with aldehydes and ketones*, in *Org. React.*, 1954, **8**, 364.

S. D. Burke and P. A. Grieco, *Intramolecular reactions of diazocarbonyl compounds*, in *Org. React.*, 1979, **26**, 361.

H. E. Simmons, T. L. Cairns, S. A. Vladuchick and C. M. Hoiness, *Cyclopropanes from unsaturated compounds, methylene iodide and zinc:copper couple*, in *Org. React.*, 1973, **20**, 1.

V. Dave and E. W. Warnhoff, *Reactions of diazoacetic esters with alkenes, alkynes, heterocyclic and aromatic compounds*, in *Org. React.*, 1970, **18**, 217.

M. Schuster and S. Blechert, *Olefin metathesis in organic chemistry*, in *Angew. Chem. Int. Ed. Engl.*, 1997, **36**, 2036.

M. Kloetzel, *The Diels–Alder reaction with maleic anhydride*, in *Org. React.*, 1948, **4**, 1.

H. L. Holmes, *The Diels–Alder reaction, ethylenic and acetylenic dienophiles*, in *Org. React.*, 1948, **4**, 60.

L. L. Butz and A. Rytina, *The Diels–Alder reaction, quinones and other cyclic enones*, in *Org. React.*, 1949, **5**, 136.

E. Ciganek, *The intramolecular Diels–Alder reaction*, in *Org. React.*, 1984, **32**, 1.

S. Blechert, *The hetero-Cope rearrangement in organic synthesis*, in *Synthesis*, 1989, 71.

6

Methods of Making Carbon–Nitrogen Bonds

Aims

The aim of this chapter is to describe methods of making carbon–nitrogen bonds. By the end of this chapter should understand the application of:

- Nucleophilic nitrogen species in substitution and addition reactions in making carbon–nitrogen bonds
- Electrophilic nitrogen species in making carbon–nitrogen bonds
- Rearrangement reactions in making carbon–nitrogen bonds
- The interconversion of different nitrogen-containing functional groups in synthesis

6.1 Introduction

Methods for making carbon–nitrogen bonds use reagents containing electrophilic, nucleophilic, radical, nitrene and dipolar nitrogen species. Some examples are given in Table 6.1.

These reactive nitrogen species can participate in substitution and addition reactions. New carbon–nitrogen bonds can also be formed by

Table 6.1 Examples of reactive nitrogen species

Electrophilic nitrogen	NO^+, NO_2^+, ArN_2^+
Nucleophilic nitrogen	NH_3, NH_2^-
Radical nitrogen	NO·
Dipolar nitrogen	$\bar{C}H_2-\overset{+}{N}\equiv N$
Nitrenes	$R_3C-N:$

a series of rearrangement reactions. Once a carbon–nitrogen bond has been made, the oxidation level of the nitrogen function may be changed. For example, a nitro group or an imine can be reduced to an amine. Hence it is possible to introduce a nitrogen function by using an electrophilic species (*e.g.* a nitronium ion), even though the final objective may be a functional group such as an amine with a nucleophilic nitrogen.

6.2 Electrophilic Methods of Making C–N Bonds

Reactants containing an electrophilic nitrogen atom are used for making carbon–nitrogen bonds to electron-rich species such as alkenes and arenes. The electrophilic **nitronium ion** (NO_2^+), which is used in aromatic nitrations, can be generated from concentrated nitric acid by the action of concentrated sulfuric acid (the classical **mixed acid**) or of acetic acid and/or acetic anhydride. The latter may involve the formation of the potentially explosive acetyl nitrate. Other methods have used nitric acid in the presence of a Lewis acid catalyst such as boron trifluoride, lanthanide(III) trifluoromethanesulfonates or a zeolite. Nitronium tetrafluoroborate ($NO_2^+BF_4^-$) is a useful reagent for the nitration of acid-sensitive aromatic compounds such as aromatic nitriles. Some metal nitrates such as copper(II) nitrate or vanadium(V) oxytrinitrate have been employed in nitration. The combination of ozone and nitrogen dioxide (the **Kyodai nitration**) has provided a relatively non-acidic methodology based on the formation of N_2O_5 and its fission to give $NO_2^+NO_3^-$.

A number of nitrations, particularly of phenols, may be nitrosations followed by oxidation of the nitro compound.

Although the underlying pattern of aromatic nitration (*ortho*/*para* or *meta*) is determined by the substituents on the aromatic ring, the ratio of the different isomeric compounds that are formed may vary with the reagent. For example, nitrations using a zeolite catalyst can show a high *para* selectivity whilst the corresponding Kyodai nitration can give a significant proportion of the *ortho* isomers.

The conditions required for the nitration of some typical aromatic compounds are given in Table 6.2.

Although the nitration of aromatic rings that are activated by electron-donating substituents (*e.g.* methoxy groups) takes place at the *ortho* and *para* positions under relatively mild conditions, the nitration of compounds that contain electron-withdrawing groups requires more vigorous conditions.

The **nitration of alkenes** by nitric acid involves an electrophilic addition of the nitronium ion. It is often accompanied by the loss of a proton to give a nitroalkene.

$$Ph_2C{=}CH_2 \xrightarrow{HNO_3} Ph_2C{=}CHNO_2$$

Table 6.2 Conditions for the nitration of aromatic compounds

Substrate	*Conditions*	*Major product*
Benzene	Conc. HNO_3/conc. H_2SO_4, 50 °C, 1 h	Nitrobenzene
Acetanilide	Conc. HNO_3/conc. H_2SO_4, 0 °C, then 30 min at room temp.	4-Nitroacetanilide
1,2-Dimethoxybenzene	Conc. HNO_3/water (1:1) in glacial acetic acid	1,2-Dimethoxy-4-nitrobenzene
Benzoic acid	Conc. HNO_3/conc. H_2SO_4 at 60 °C for 1 h	3-Nitrobenzoic acid
Nitrobenzene	Conc. HNO_3/conc. H_2SO_4 at 60 °C for 1 h	1,3-Dinitrobenzene

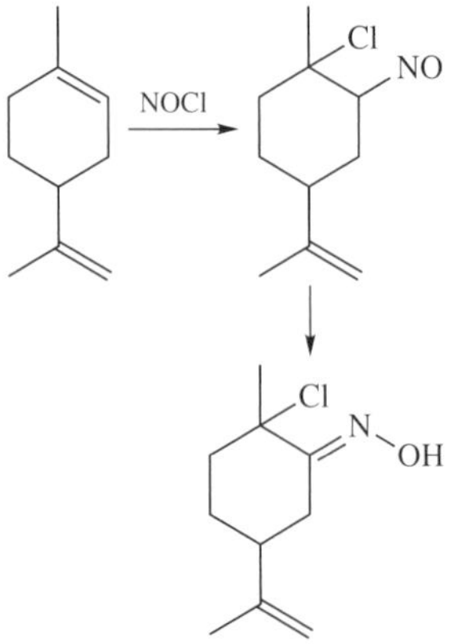

There are a number of useful reactions of the **nitrosonium ion** (NO^+) which lead to the formation of new carbon–nitrogen bonds. In this context it is important to note that aliphatic *C*-nitroso compounds ($R_2CH–N=O$) are tautomeric with oximes ($R_2C=N–O–H$). The **nitrosation** of phenols with nitrous acid, generated from sodium nitrite and sulfuric acid, takes place very easily to give monosubstituted compounds. Nitrosyl chloride (NOCl) is an effective nitrosating agent for alkenes, giving nitrosochlorides.

The nitrosation of the position adjacent to a carbonyl group is a useful way of introducing a nitrogen function by giving an α-oximino ketone. The reaction is that of the enol.

The **coupling of diazonium salts** with phenols leads to the formation of new carbon–nitrogen bonds. Synthetic uses of this include the reduction of the diazo group to a hydrazone.

6.3 Nucleophilic Methods of Forming C–N Bonds

Reactions involving nucleophilic nitrogen which lead to the formation of carbon–nitrogen bonds may be classified in terms of the oxidation level of the electron-deficient centre at which the nitrogen reacts. These involve substitution reactions of alkyl halides and similar leaving groups, addition reactions to carbonium ions and addition reactions to carbonyl groups and similar electron-deficient centres.

The **nucleophilic substitution** of an alkyl halide by ammonia leads sequentially to primary, secondary, tertiary and finally quaternary ammonium salts. Since alkyl groups are electron donating, the nucleophilicity of a primary amine is greater than that of ammonia.

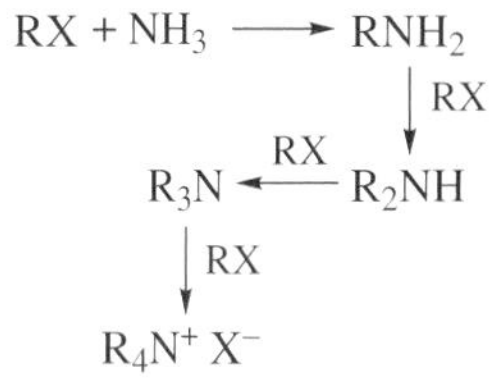

Consequently, the simple nucleophilic substitution of an alkyl halide is difficult to control in order to obtain a primary or a secondary amine. However, as the steric size of the groups attached to the nitrogen increases, so the ease of the reaction decreases.

$$RX + NH_3 \longrightarrow RNH_2 \xrightarrow{RX} R_2NH \xrightarrow{RX} R_3N \xrightarrow{RX} R_4N^+ X^-$$

A number of methods have been developed to achieve the selective formation of primary amines. When a carbonyl group is attached to the nitrogen of an amine, converting it to an amide, the nucleophilicity of the nitrogen is reduced but the acidity of the N–H is increased. Use of this is made in the **Gabriel synthesis** of primary amines. The two carbonyl groups of phthalimide render the hydrogen acidic. The potassium salt of phthalimide contains a nitrogen anion which is a powerful nucleophile and which may be used in the nucleophilic substitution of alkyl halides. Once substitution has taken place, the amide is non-basic and only one alkyl group becomes attached to the nitrogen (Scheme 6.1). Hydrolysis of the amide generates the primary amine. Hydrazine may be used for this purpose. Hydrazinolysis of the first carboxamide brings the basic terminal of the hydrazide close to the second amide to facilitate the cleavage (Scheme 6.1).

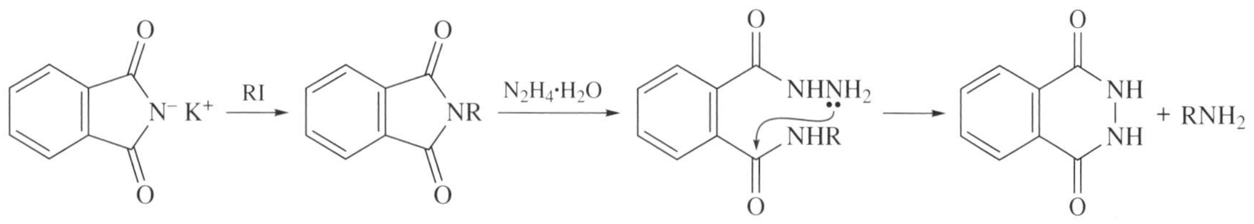

N⁻ K⁺ —(RI)→ NR —($N_2H_4{\cdot}H_2O$)→ NHNH₂ / NHR → NH–NH + RNH_2

Scheme 6.1 Gabriel primary amine synthesis

The electron-withdrawing trifluoromethyl group of trifluoroacetamide not only increases the acidity of the amide N–H so that the trifluoroacetamide anion can be used as a nucleophile in this type of reaction, but it also helps the hydrolysis of the alkylated amide. The method has been extended to the preparation of secondary amines through the dialkylation of sulfonamides (Scheme 6.2).

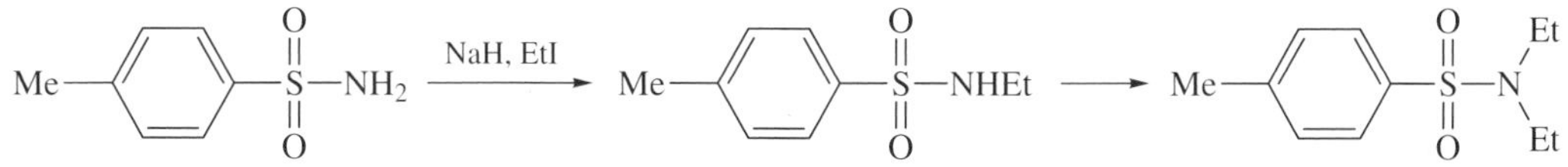

$$Me{-}C_6H_4{-}SO_2{-}NH_2 \xrightarrow{NaH,\ EtI} Me{-}C_6H_4{-}SO_2{-}NHEt \longrightarrow Me{-}C_6H_4{-}SO_2{-}NEt_2$$

Scheme 6.2

Sodium azide is a useful nucleophile which can be used in substitution reactions of alkyl halides or methanesulfonates. The azide will only react once to give a monosubstituted compound. In a second step, catalytic reduction with hydrogen over a palladium on charcoal catalyst leads to a primary amine (Scheme 6.3).

$$\underset{\displaystyle Br}{MeCHCO_2H} \xrightarrow{NaN_3} \underset{\displaystyle N_3}{MeCHCO_2H} \xrightarrow{H_2,\ Pd} \underset{\displaystyle NH_2}{MeCHCO_2H}$$

Scheme 6.3

A secondary amine can be prepared by making use of the fact that a nitroso group on an aromatic ring activates a substituent in the *para* position to nucleophilic substitution. Nitrosation of a dialkylaniline gives the *p*-nitrosodialkylaniline which, on treatment with alkali, gives *p*-nitrosophenol and the dialkylamine (Scheme 6.4).

$$C_6H_5NH_2 \xrightarrow{RI} C_6H_5NR_2 \xrightarrow{HNO_2} p\text{-}ON{-}C_6H_4{-}NR_2 \xrightarrow{NaOH} p\text{-}ON{-}C_6H_4{-}OH + R_2NH$$

Scheme 6.4

New carbon–nitrogen bonds can be obtained by the nucleophilic addition of an amino nitrogen derivative to a carbonyl group. The products of these additions readily lose water to form imines. The addition to acid derivatives such as acyl chlorides or anhydrides yields amides. Reduction of amides, imines and oximes with sodium in ethanol, lithium aluminium hydride, sodium borohydride or, in some instances, sodium cyanoborohydride, provides a useful way of making amines (Scheme 6.5).

The reductive amination of ketones by the **Leuckart method** uses ammonium formate both as the source of nitrogen to form an imine and as the reductant (Scheme 6.5). Another way of achieving this is to react the ketone with ammonium acetate and sodium cyanoborohydride.

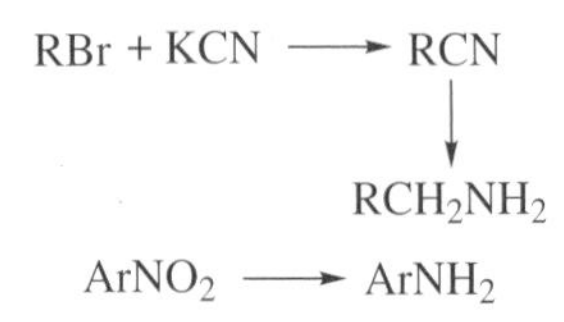

The synthetic value of reductive methods lies in the fact that these reactions allow the nitrogen to be introduced into a compound by a variety of different methods at different oxidation levels. Nitriles may be introduced by nucleophilic substitution or addition reactions, while nitro compounds may be prepared by electrophilic aromatic nitrations. Reduction of the nitro compounds by iron or tin and hydrochloric or sulfuric acid gives the primary amines as their salts. Nitriles can be reduced with lithium aluminium hydride.

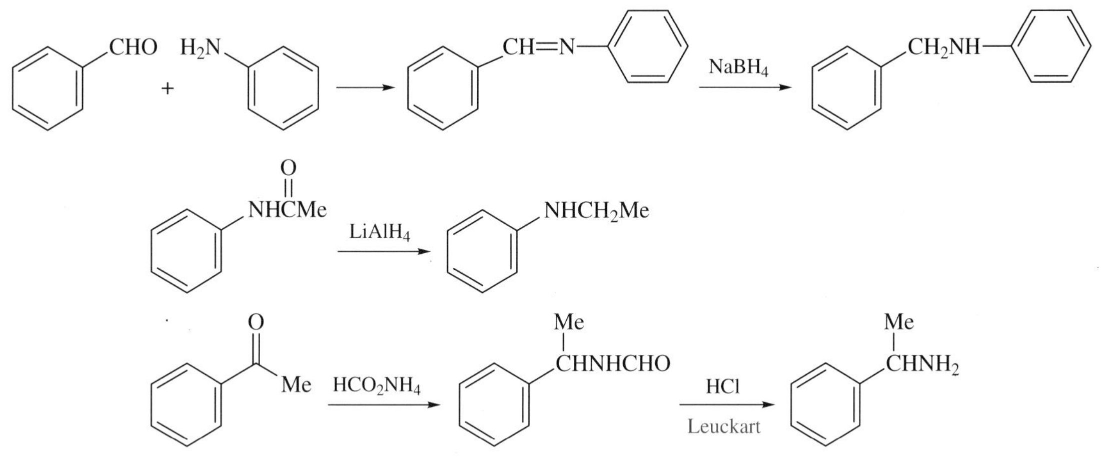

Scheme 6.5 Amine syntheses, including the Leuckart reaction

Worked Problem 6.1

Q Propose a synthesis of tropinone (**1**).

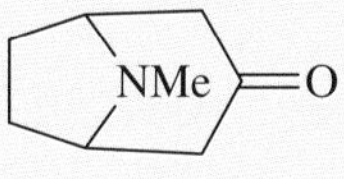

A Analysis of the target molecule reveals a β-amino ketone, which is the product of a Mannich reaction. The Mannich reaction involves the condensation of an enolate with an iminium salt. The target molecule contains a propanone unit. Activation of this as 3-oxopentanedioic acid (acetone dicarboxylic acid) provides a more effective enolate unit. This synthesis by Sir Robert Robinson was one of the classical syntheses of the 20th century.

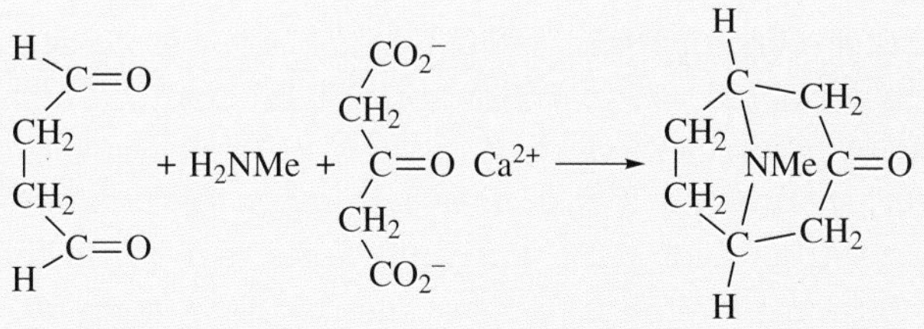

6.4 Rearrangement Methods

There is a series of methods that are used to make new carbon–nitrogen bonds which involve rearrangements. The **Beckmann rearrangement** of oximes with sulfuric acid gives amides, which may be hydrolysed or reduced to form amines. The stereochemistry of the reaction involves a *trans* relationship between the hydroxyl group of the oxime and the carbon–carbon bond, which undergoes the 1,2-shift. This defines the structure of the amide that is formed, and hence the amine that is obtained on hydrolysis (Scheme 6.6).

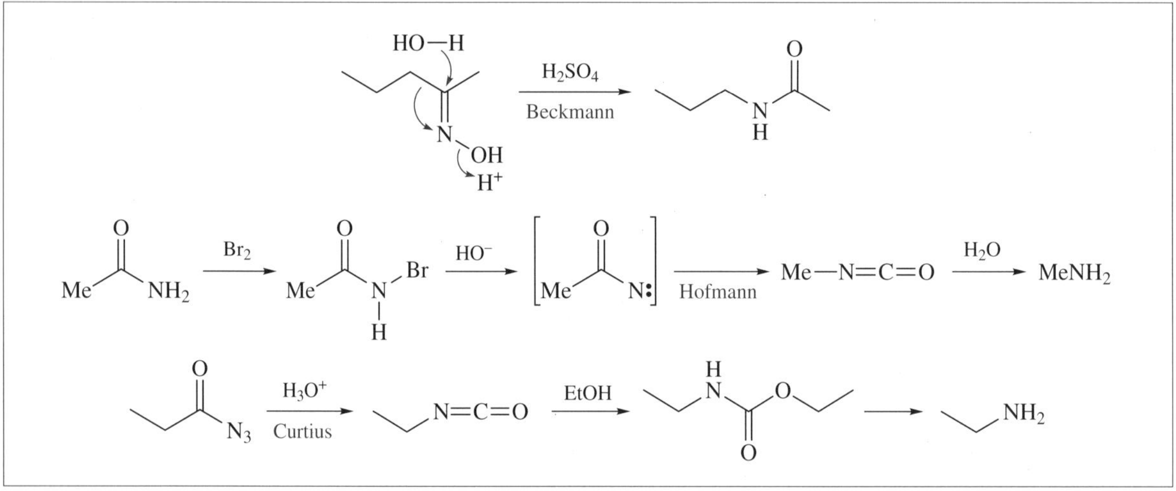

HO—H; N; OH; H^+ → (H_2SO_4, Beckmann) → O; N H

Me(C=O)NH$_2$ → (Br_2) → Me(C=O)N(H)Br → (HO^-) → [Me(C=O)N:] → (Hofmann) → Me—N=C=O → (H_2O) → $MeNH_2$

(C=O)N_3 → (H_3O^+, Curtius) → N=C=O → (EtOH) → H N O O → NH_2

Scheme 6.6 Beckmann, Hofmann and Curtius rearrangements

A number of rearrangements of derivatives of carboxylic acids involve the formation of nitrene intermediates. The best known of these is the **Hofmann degradation** of amides. The amide is treated with bromine and alkali to give the *N*-bromo compound. The compound undergoes an α-elimination to form a nitrene intermediate, which immediately rearranges to an isocyanate. Hydrolysis of the isocyanate generates the amine (Scheme 6.6).

The **Curtius rearrangement** of acid azides and the **Lossen rearrangement** of hydroxamic acids follow a similar pattern. The Curtius rearrangement involves the loss of nitrogen from the azide to form the nitrene intermediate and thence an isocyanate. The latter may be converted directly to an amine or indirectly *via* a urethane (Scheme 6.6). In these degradations of acid derivatives the carboxyl group of the acid is lost and the amine which is formed has one less carbon atom.

6.5 The Synthesis of Amino Acids

The synthesis of α-amino acids illustrates a number of the general points of synthesis which have been discussed in previous chapters. The displacement of halogen from α-halogeno acids can be used. The zwitterionic character of the products reduces the extent of further substitution by the amine nitrogen.

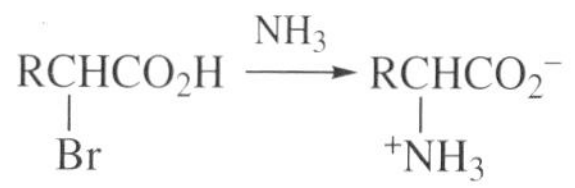

$RCH(Br)CO_2H \xrightarrow{NH_3} RCH(^{+}NH_3)CO_2^{-}$

The **Strecker synthesis** of amino acids and a related method involving a hydantoin (Scheme 6.7) use addition to a carbonyl group, possibly through the addition of a cyanide ion to an imine. The **Erlenmeyer azlactone** (oxazolone) **synthesis** involves a condensation reaction of glycine. The condensation reaction requires an acidic C–H and hence the other more acidic hydrogen atoms of glycine must be removed. In order to do this, glycine is converted first to benzoylglycine (hippuric acid), which then undergoes a cyclodehydration to form an oxazolone (azlactone) (Scheme 6.8). In this intermediate the only acidic hydrogen atoms are those adjacent to the carbonyl group. Condensation with an aldehyde, such as benzaldehyde, may take place. In this case, reduction and hydrolysis then affords the amino acid, phenylalanine.

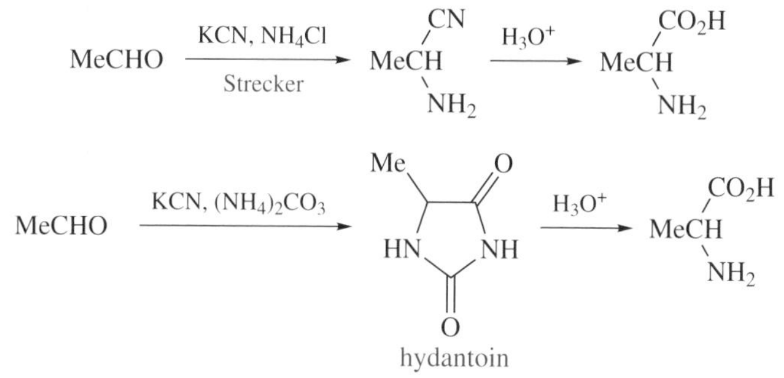

$MeCHO \xrightarrow[\text{Strecker}]{KCN,\ NH_4Cl} MeCH(CN)NH_2 \xrightarrow{H_3O^+} MeCH(CO_2H)NH_2$

$MeCHO \xrightarrow{KCN,\ (NH_4)_2CO_3}$ hydantoin $\xrightarrow{H_3O^+} MeCH(CO_2H)NH_2$

Scheme 6.7 Strecker amino acid synthesis

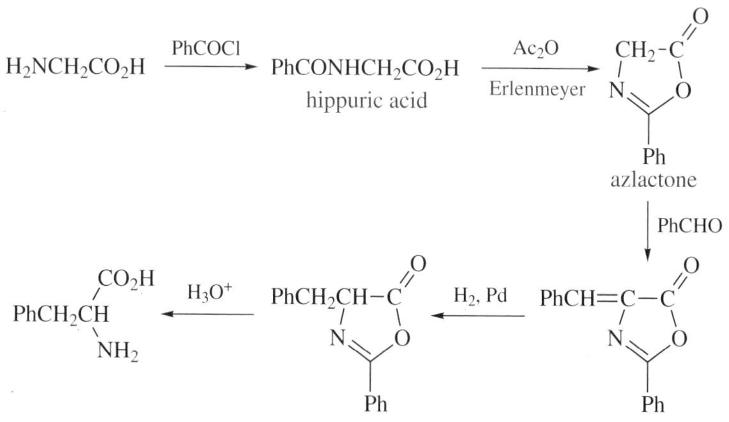

$H_2NCH_2CO_2H \xrightarrow{PhCOCl} PhCONHCH_2CO_2H$ (hippuric acid) $\xrightarrow[\text{Erlenmeyer}]{Ac_2O}$ azlactone $\xrightarrow{PhCHO} PhCH{=}C$ (azlactone) $\xrightarrow{H_2,\ Pd} PhCH_2CH$ (azlactone) $\xrightarrow{H_3O^+} PhCH_2CH(CO_2H)NH_2$

Scheme 6.8 Erlenmeyer azlactone synthesis

Two methods based on diethyl malonate (propanedioate) have been used. In the first, the bromine of diethyl bromomalonate is displaced by the potassium salt of phthalimide to introduce the nitrogen. The remaining acidic hydrogen, which is activated by both carbonyl groups, may then be alkylated. Hydrolysis and decarboxylation afford the amino acid (Scheme 6.9). However, the most frequently employed procedure is that involving an acetamidomalonate. Nitrosation of diethyl malonate gives the oximinomalonate, which is reduced with zinc and acetic acid to form the acetamidomalonate (Scheme 6.10). This may be alkylated, hydrolysed and decarboxylated to give the amino acid (Scheme 6.10).

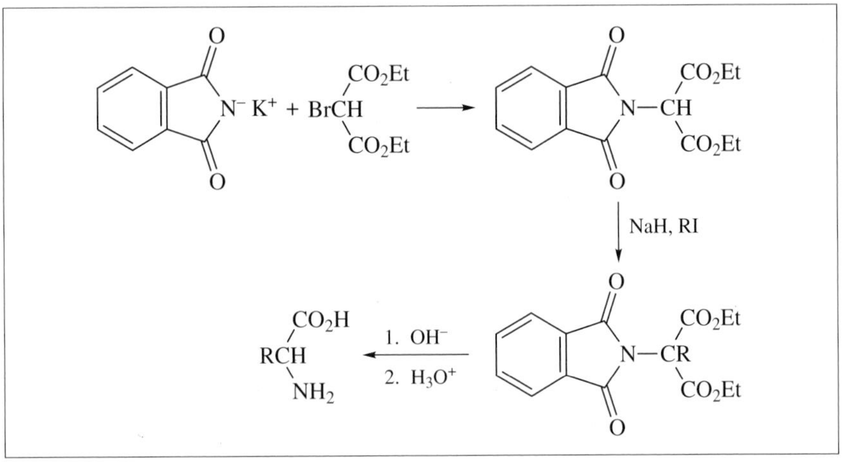

Phthalimide $N^- K^+$ + $BrCH(CO_2Et)_2$ ⟶ phthalimido-$N{-}CH(CO_2Et)_2$ $\xrightarrow{\text{NaH, RI}}$ phthalimido-$N{-}CR(CO_2Et)_2$ $\xrightarrow{\text{1. OH}^-,\ \text{2. H}_3\text{O}^+}$ $RCH(CO_2H)(NH_2)$

Scheme 6.9

$$CH_2(CO_2Et)_2 \xrightarrow[H_3O^+]{NaNO_2,} HO{-}N{=}C(CO_2Et)_2 \xrightarrow[AcOH]{Zn,} AcNH{-}CH(CO_2Et)_2 \xrightarrow[NaOEt]{RI,} AcNH{-}CR(CO_2Et)_2 \xrightarrow[2.\ H_3O^+]{1.\ OH^-} RCH(CO_2H)(NH_2)$$

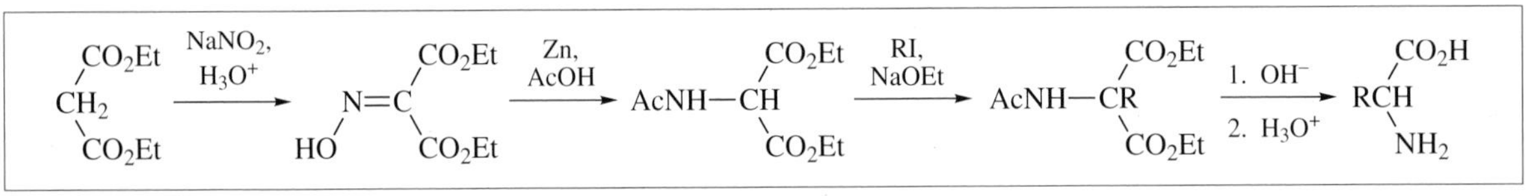

Scheme 6.10

The construction of peptides from their constituent amino acids involves the formation of an amide bond between the amino group of one amino acid and the carboxyl group of another. The principles of protection and activation required to generate the correct peptide bond are discussed in Chapter 8.

Worked Problem 6.2

Q Propose a synthesis of [3-^{14}C]valine (**2**).

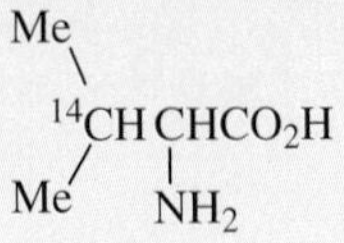

$$Me_2{}^{14}CH{-}CH(NH_2)CO_2H \quad \mathbf{2}$$

A The acetamidomalonate route is a useful general route for the synthesis of amino acids. Alkylation with 2-bromo[2-^{14}C]propane, followed by hydrolysis and decarboxylation, gives the labelled amino acid:

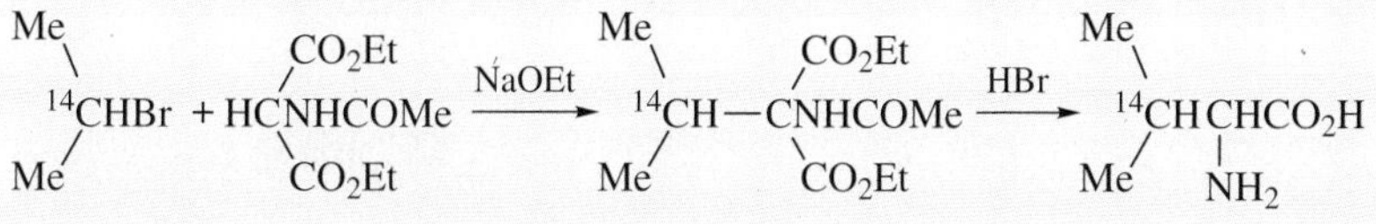

$$Me_2{}^{14}CHBr + HC(CO_2Et)_2NHCOMe \xrightarrow{NaOEt} Me_2{}^{14}CH{-}C(CO_2Et)_2NHCOMe \xrightarrow{HBr} Me_2{}^{14}CHCH(NH_2)CO_2H$$

6.6 The Synthesis of Nitrogen Heterocycles

Many biologically active molecules contain a heteroaromatic ring, particularly one which includes nitrogen. The synthesis of these illustrates many of the principles of carbon–nitrogen bond formation. The nitrogen may be present either as an **imine** (C=N) or as an **enamine** (C=C–N–). Both may arise from a **carbonyl–amine condensation** reaction. This relationship is helpful in understanding the synthesis of heteroaromatic compounds containing nitrogen (Scheme 6.11).

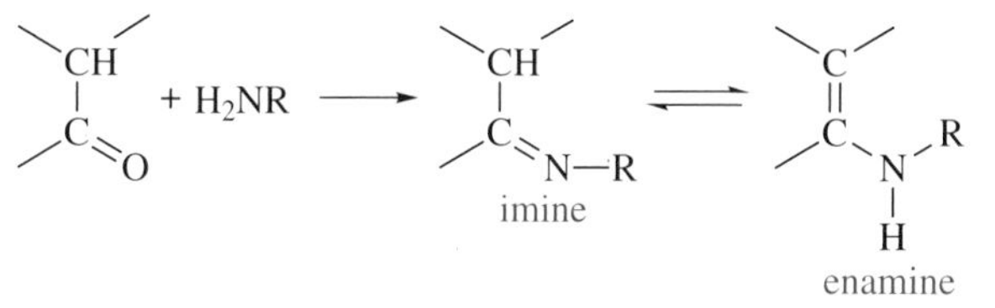

$$>CH{-}C(=O){-} + H_2NR \longrightarrow >CH{-}C(=N{-}R){-} \rightleftharpoons >C{=}C(NHR){-}$$

imine   enamine

Scheme 6.11

The synthesis of many heteroaromatic compounds may be divided into the formation of the appropriate carbonyl component and the condensation with a nucleophilic component containing the heteroatom. This is illustrated by the condensation of hexane-2,5-dione (Scheme 6.12) with ammonia to give 2,5-dimethylpyrrole and the condensation of pentane-2,4-dione (Scheme 6.12) with hydrazine to give 3,5-dimethylpyrazole.

Scheme 6.12

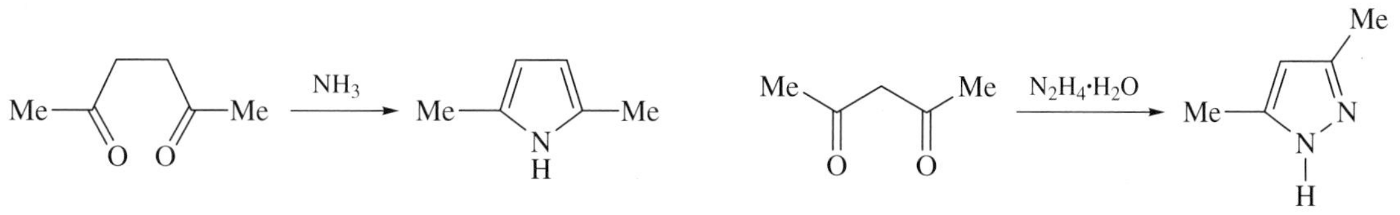

$$MeCOCH_2CH_2COMe \xrightarrow{NH_3} \text{2,5-dimethylpyrrole (Me, Me, N–H)}$$

$$MeCOCH_2COMe \xrightarrow{N_2H_4\cdot H_2O} \text{3,5-dimethylpyrazole (Me, Me, N, N–H)}$$

Not all heteroaromatic syntheses involve the reactions of nucleophilic nitrogen. Some involve diazonium chemistry and others involve pericyclic reactions. A useful classification of heteroaromatic syntheses is to consider them in terms of the reactivity of the heteroatom component.

A number of heteroatom components have two sites of reactivity. Some typical bidentate nucleophiles are shown in Box 6.1

Box 6.1 Bidentate Nucleophiles

Class A–B

Example H_2NNH_2 $O{=}C(NH_2)_2$

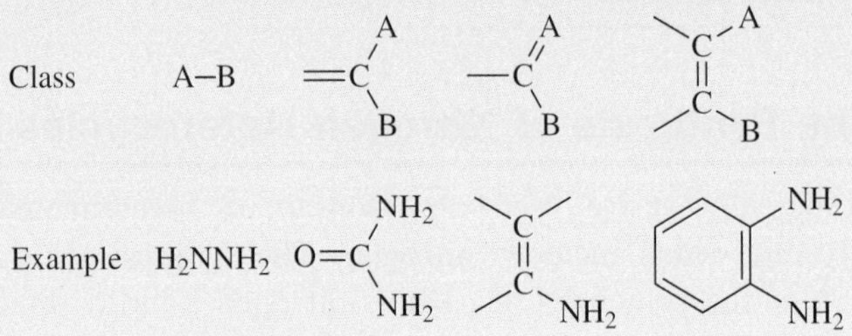

Urea is a useful diamide component, and this is exemplified by its reaction with diethyl malonate which gives the pyrimidine barbituric acid (Scheme 6.13).

barbituric acid

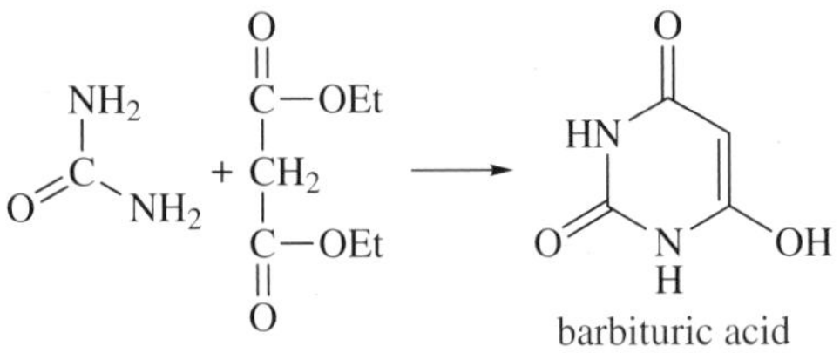

Scheme 6.13

The sulfur of thiourea is quite nucleophilic and will displace halogen from an alkyl halide. For example, treatment of chloropropanone with thiourea gives a thiazole (Scheme 6.14).

Enamines play an important role in a number of heteroaromatic syntheses. The **Hantzsch synthesis of dihydropyridines** involves condensation between a β-keto ester (or β-diketone), an aldehyde and ammonia

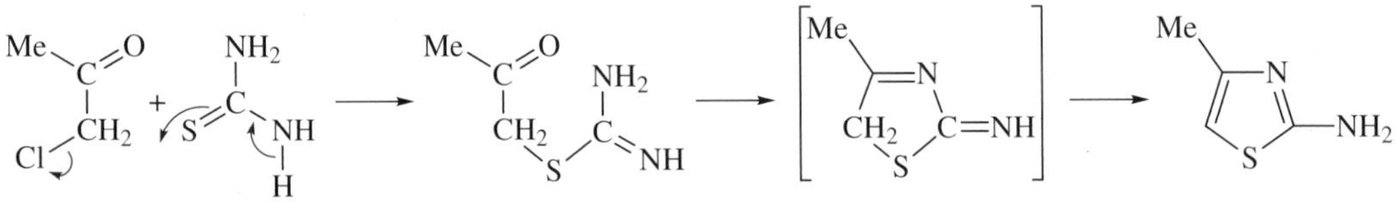

Scheme 6.14

in one vessel. The condensation between the aldehyde and the β-keto ester forms an unsaturated ketone. Ammonia reacts with a second molecule of the β-keto ester to form an enamine. A Michael-type of addition of these components leads to the carbon skeleton of the dihydropyridine. The final cyclization involves a carbonyl–amine condensation between the enamine and the more electron-deficient carbonyl group of the β-keto ester. Pyridines may be obtained by a subsequent oxidation of the dihydropyridine (Scheme 6.15).

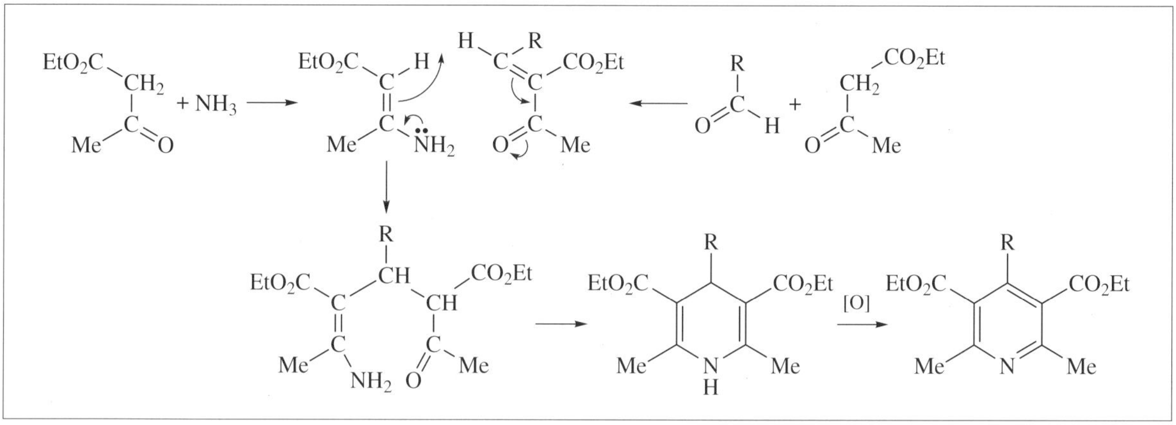

Scheme 6.15 Hantzsch dihydropyridine synthesis

The **Hantzsch synthesis of pyrroles** follows a similar sequence. The reaction involves the condensation between a β-keto ester and an α-chloro ketone in the presence of ammonia. An enamine is formed in the reaction between the β-keto ester and ammonia. The enamine displaces the chlorine of the α-chloro ketone and this is followed by the condensation with the amine to form a pyrrole (Scheme 6.16).

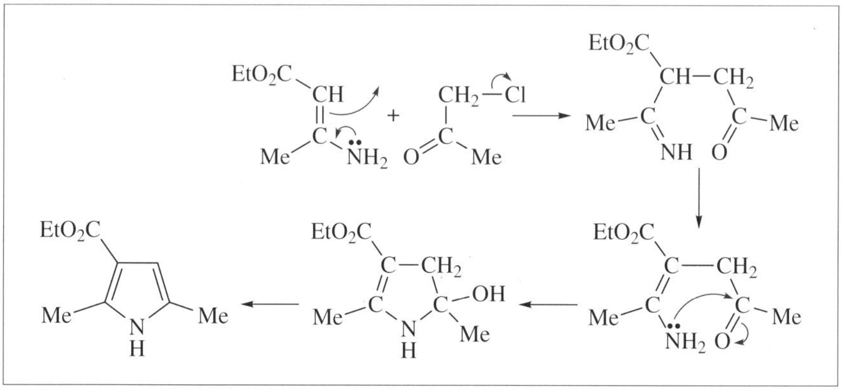

Scheme 6.16 Hantzsch pyrrole synthesis

The **Knorr synthesis of pyrroles** involves a combination of a carbanion condensation to form the C–C bond and a carbonyl–amine condensation to introduce the heteroatom (Scheme 6.17).

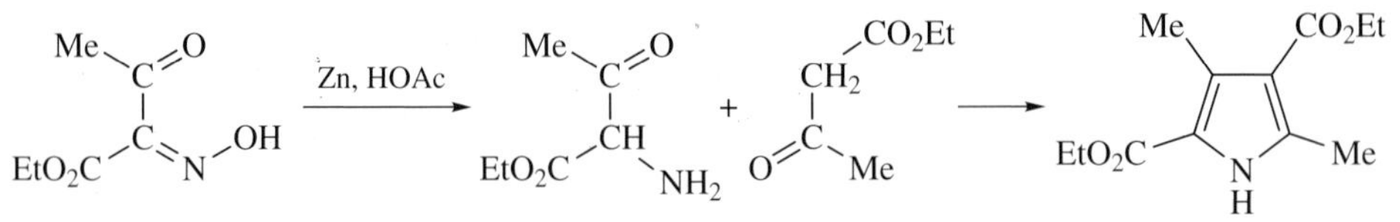

Scheme 6.17 Knorr pyrrole synthesis

The syntheses of quinolines from aniline also fall into this pattern. Both the **Skraup** and **Doebner–von Miller syntheses** rely on the formation of an unsaturated aldehyde, which undergoes a Michael addition of aniline followed by an acid-catalysed cyclization (Scheme 6.18).

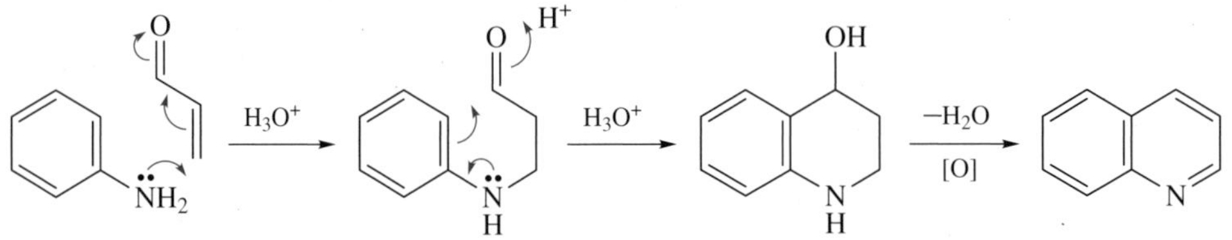

Scheme 6.18 Skraup synthesis

Diamines such as *o*-phenylenediamine provide the hetero component for a number of syntheses. Thus, condensation with formic acid gives a benzimidazole, while condensation with butane-2,3-dione gives dimethylquinoxaline (Scheme 6.19).

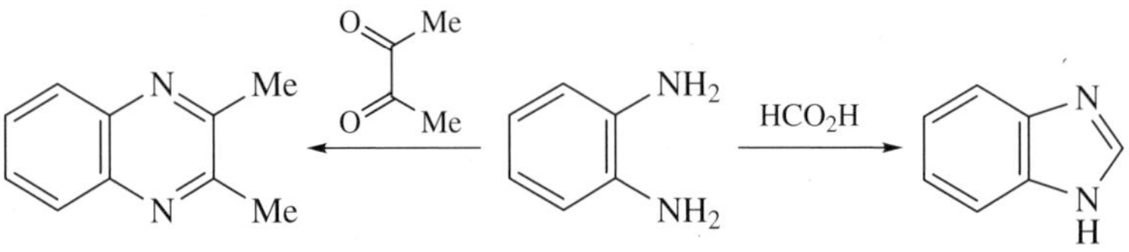

Scheme 6.19

Anthranilic acid illustrates another family of hetero components which contain both an electron-deficient carbon and a nucleophilic nitrogen. The cyclization of *N*-acylated anthranilic acids to benzoxazines (Scheme 6.20) is an example.

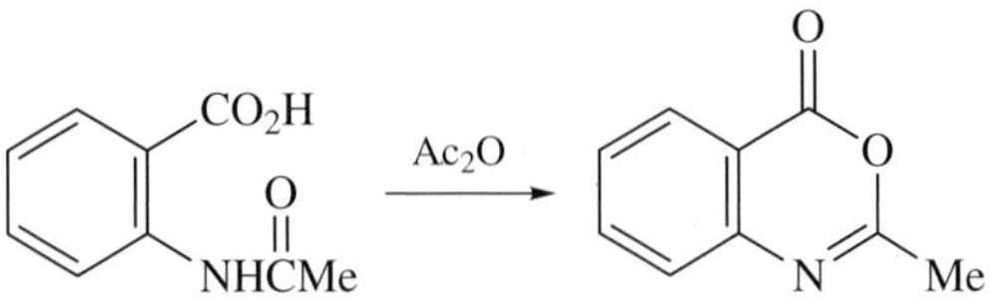

Scheme 6.20

A completely different mechanism is involved in the **Fischer indole** synthesis. The reaction involves an electrocyclic rearrangement of a phenylhydrazone (Scheme 6.21).

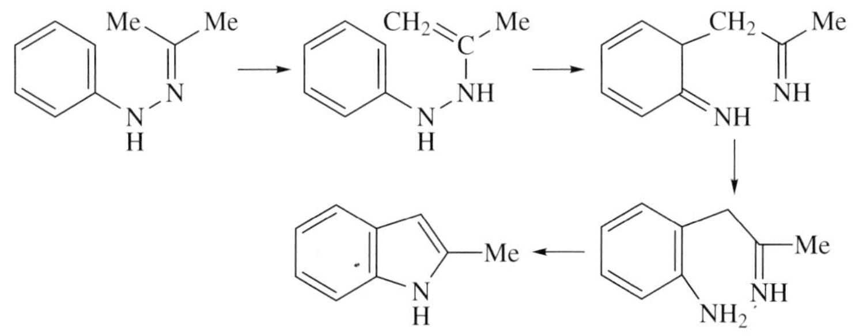

Scheme 6.21 Fischer indole synthesis

Worked Problem 6.3

Q Propose a synthesis of the substituted imidazole **3**.

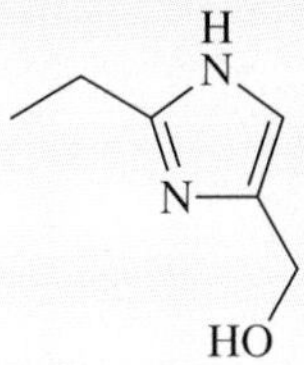

A Routes for the construction of a nitrogen heterocycle may be revealed by mentally replacing the nitrogen atom by an oxygen atom and then considering the synthesis of the hydroxyl or carbonyl components which may be revealed. There are several possible dissections of the imidazole ring using this strategy. A fruitful dissection is as follows:

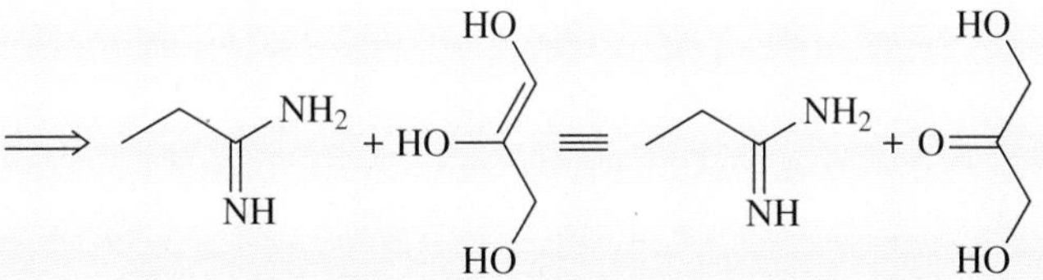

This suggests a synthesis involving the condensation between an amidine and 2-oxopropane-1,3-diol (dihydroxyacetone):

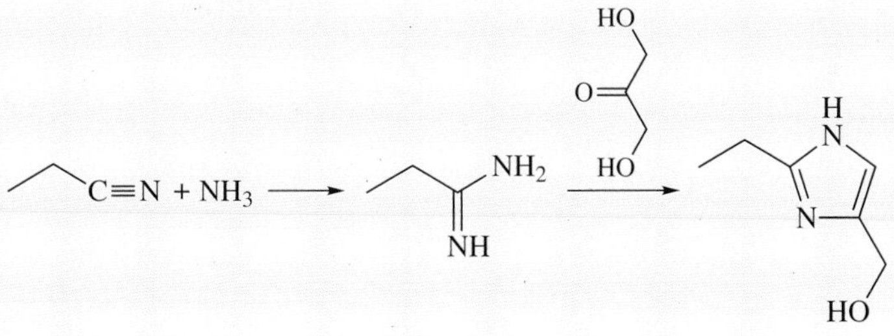

Worked Problem 6.4

Q Propose a synthesis of the antibacterial sulfonamide sulfadimidine (**4**).

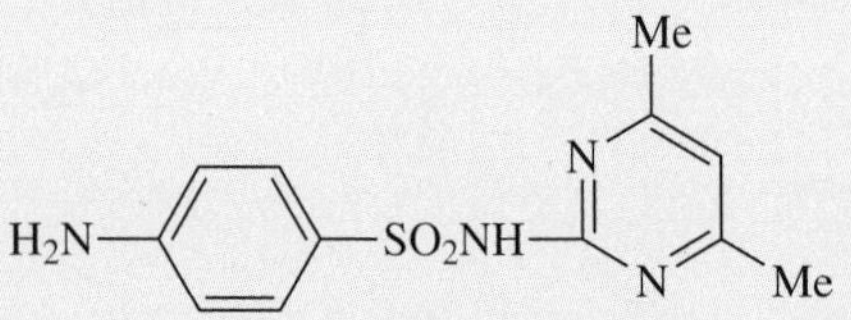

A This synthesis reveals the importance of not just looking for strategic bonds in the target molecule but also of dissecting the structure in terms of potential building blocks. Examination of the target molecule reveals a number of building blocks:

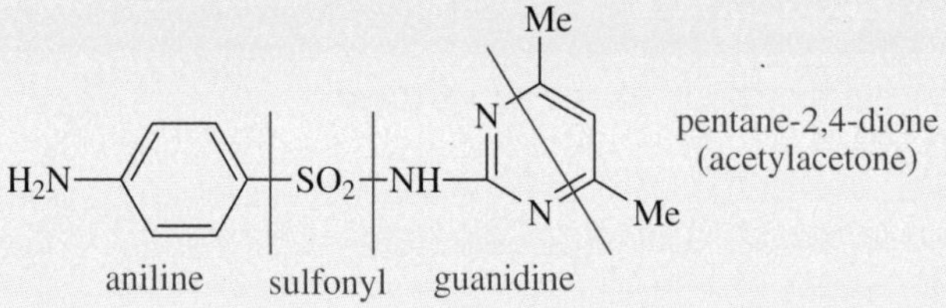

If the nitrogen atoms of the pyrimidine ring are replaced by oxygen, the pentane-2,4-dione (acetylacetone) fragment becomes clear. The identification of these building blocks suggests a synthetic sequence starting by chlorosulfonation of acetanilide:

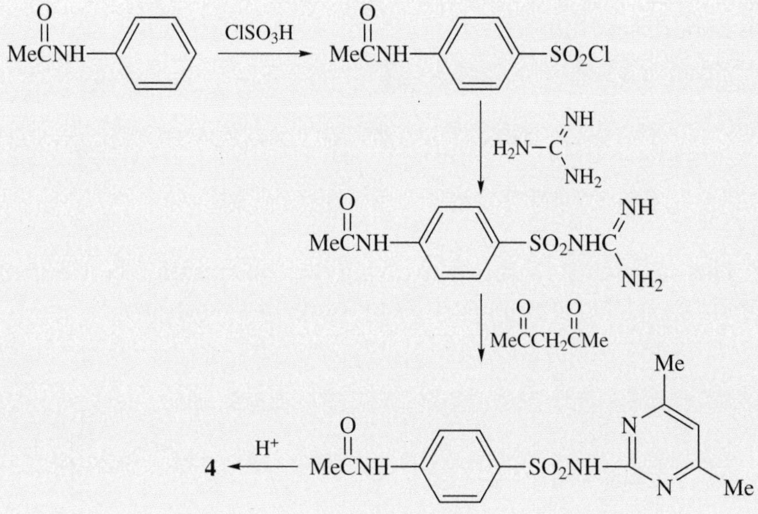

Summary of Key Points

1. Methods of making C–N bonds may be divided into electrophilic, nucleophilic, radical, dipolar and nitrene and rearrangement methods. Functional group interconversions (*e.g.* reduction of nitro groups to amines) are important.

2. Electrophilic nitration is the major method for making aromatic C–N bonds.

3. Nucleophilic methods may be divided into substitution and carbonyl addition methods.

4. The simple nucleophilic substitution of alkyl halides by ammonia leads to a complex mixture of primary, secondary and tertiary amines. Special methods, *e.g.* based on phthalimide, have been devised to prepare primary amines.

5. Useful amino acid syntheses are based on azlactone and diethyl acetamidomalonate chemistry.

6. Many heterocyclic syntheses involve carbonyl–amine condensation reactions. The synthesis may then be divided into the steps that lead to the carbonyl component and those that form the heterocyclic ring.

Problems

6.1. Show how the heterocyclic ring might be created in the following compounds:

(a) O Me O Me Me Me N Me H

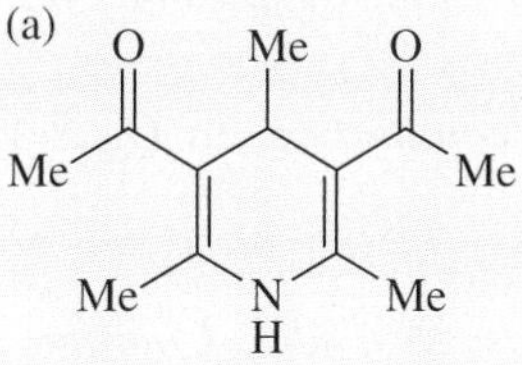

(b) EtO_2C Me Me S CO_2Et

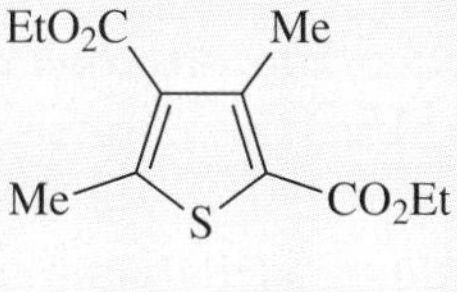

(c) O HN O N Me H

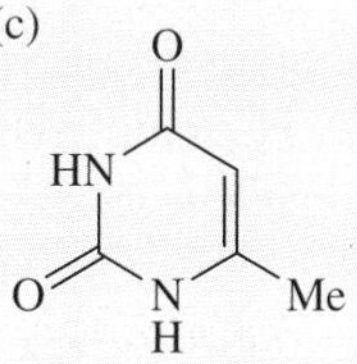

(d) N Me N Me

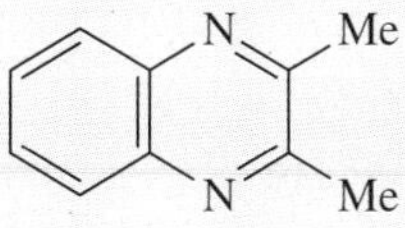

(e) MeO N Me

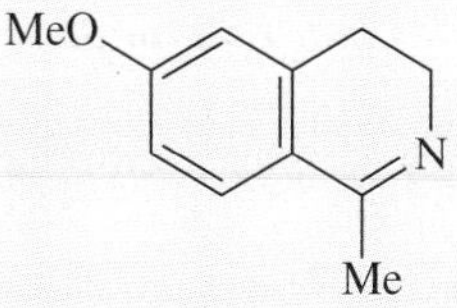

(f) Me N S CO_2Et

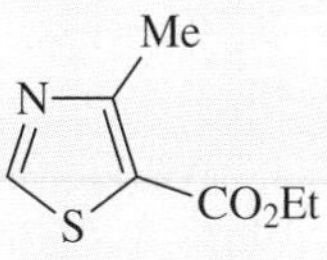

6.2. Indicate the reagents and other reactants in the following synthesis:

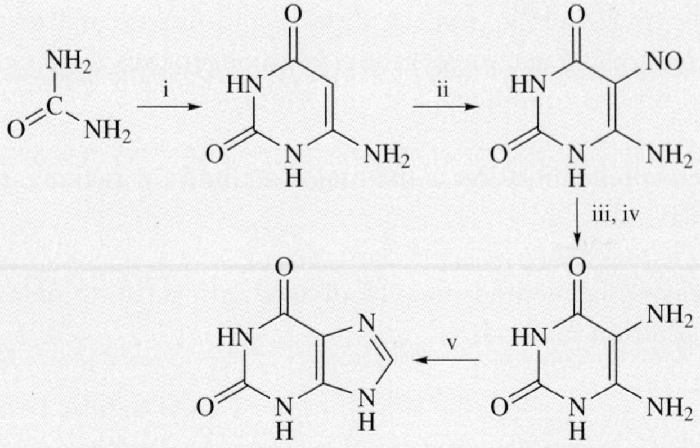

NH_2–C(=O)–NH_2 —i→ (HN, O, N-H, NH_2 pyrimidinone) —ii→ (HN, O, NO, N-H, NH_2) —iii, iv→ (HN, O, NH_2, N-H, NH_2) —v→ (HN, O, N, N-H, N-H purine)

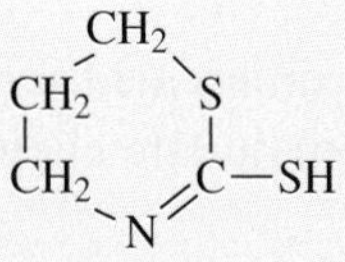

6.3. How might [1-^{14}C]phenylalanine (**5**) be prepared?

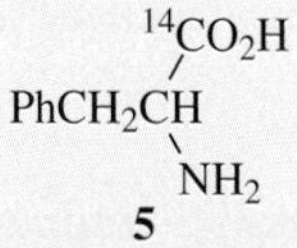

5: $PhCH_2CH(^{14}CO_2H)NH_2$

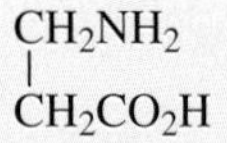

6: CH_2NH_2–CH_2CO_2H

7: CH_2, CH_2, S, CH_2, C–SH, N (ring)

6.4. How might β-alanine (**6**) be prepared from succinic acid?

6.5. How might compound **7** be prepared from 1,3-dibromopropane?

Further Reading

T. Mori and H. Suzuki, *Ozone mediated nitration of organic compounds*, in *Synlett*, 1995, 383.

J. H. Ridd, *Nitrosation, diazotization and deamination*, in *Q. Rev. Chem. Soc.*, 1961, **15**, 418.

K. Smith, A. Musson and G. A. de Boos, *Superior methodology for the nitration of simple aromatic compounds*, in *Chem. Commun.*, 1996, 464.

R. E. Gawley, *The Beckmann rearrangement*, in *Org. React.*, 1988, **35**, 1.

E. L. Wallis and J. F. Lane, *The Hofmann reaction*, in *Org. React.*, 1946, **3**, 267.

H. Wolf, *The Schmidt reaction*, in *Org. React.*, 1946, **3**, 307.
P. A. Smith, *The Curtius reaction*, in *Org. React.*, 1946, **3**, 337.
H. E. Carter, *Azlactones*, in *Org. React.*, 1946, **3**, 198.
J. Jones, *Amino Acid and Peptide Synthesis*, Oxford University Press, Oxford, 1992.
A. R. Katritzky, D. L. Ostercamp and T. I. Yousaf, *The mechanism of heterocyclic ring closures*, in *Tetrahedron*, 1987, **43**, 5171.

7

Functional Group Transformations

Aims

The aim of this chapter is to describe the role of oxidation, reduction, halogenation and the interconversion of aromatic functional groups in synthesis. By the end of the chapter you should understand:

- The different uses in synthesis of dehydrogenation, one- and two-electron transfer processes, and reactions involving the insertion of oxygen
- The role of hydrogenolysis, hydrogenation and the abstraction of oxygen in reductive processes
- The role of hydroboration in synthesis
- The methods of introducing halogen into a molecule using nucleophilic, electrophilic and radical reagents

7.1 Oxidation

It is possible to consider the reagents that are used for oxidation, and the key steps in the processes in which they participate, under a number of headings. These are dehydrogenation, one- and two-electron transfer processes, and the insertion of oxygen into a system.

7.1.1 Dehydrogenation Reactions

The key step in dehydrogenation reactions involves the removal of hydrogen from a molecule. The dehydrogenation of alicyclic natural products to aromatic compounds by sulfur or **selenium** provided an important method for determining the structures of these compounds (Scheme 7.1). Although it is normally easier to synthesize aromatic compounds using preformed aromatic rings, there are occasions in which the partial aromatization of polycyclic compounds is useful, for example in the conversion of a tetrahydronaphthalene or tetrahydroisoquinoline to the parent aromatic compound.

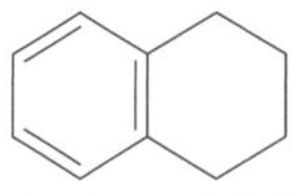

tetrahydronaphthalene

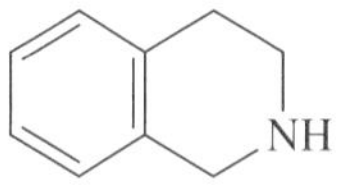

NH
tetrahydroisoquinoline

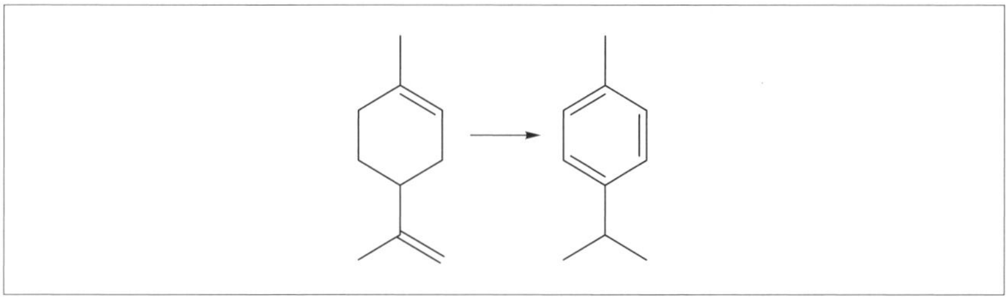

Scheme 7.1

The dehydrogenation of ketones to α,β-unsaturated ketones by quinones such as chloranil (2,3,5,6-tetrachlorobenzo-1,4-quinone) and **DDQ** (2,3-dichloro-5,6-dicyanobenzo-1,4-quinone) is a useful reaction. This may be conducted in stages to give dienones. The synthetic value of this reaction is that the electron-withdrawing role of the carbonyl group is relayed to the β-position of an α,β-unsaturated ketone, or to the δ-position of an α,β;γ,δ-dieneone. These positions then become sensitive to nucleophilic addition.

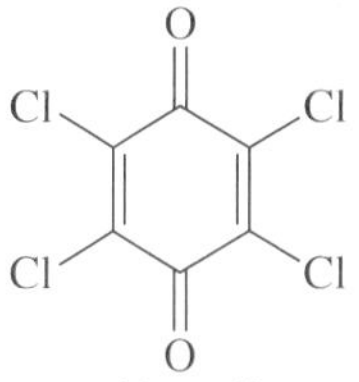

O Cl Cl Cl Cl O
chloranil

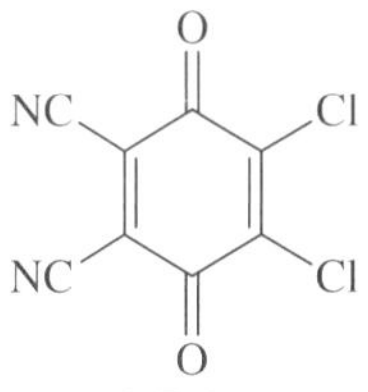

O NC Cl NC Cl O
DDQ

7.1.2 Oxidation of Alcohols

Although the oxidation of an alcohol to a ketone involves a dehydrogenation, the key step in an oxidation by a reagent such as **chromium(VI) oxide** (chromium trioxide) is a two-electron change. The first step in the reaction is the rapid formation of a chromate(VI) ester. This is followed by a fragmentation reaction in which the chromium(VI) is reduced to chromium(IV). The latter then disproportionates to give chromium(III) and chromium(VI). In simple systems such as propan-2-ol the oxidation of Me_2C^2HOH shows a marked isotope effect, confirming that the rate-determining step involves the fission of a C–H bond (Scheme 7.2). In some sterically hindered alcohols the formation of the chromate(VI) ester becomes the rate-determining feature.

A standard chromium(VI) oxide–sulfuric acid mixture (**Jones' reagent**) was originally devised for the chromium(VI) oxide oxidation of sensitive

CrO_3–H_2SO_4–H_2O
Jones' reagent

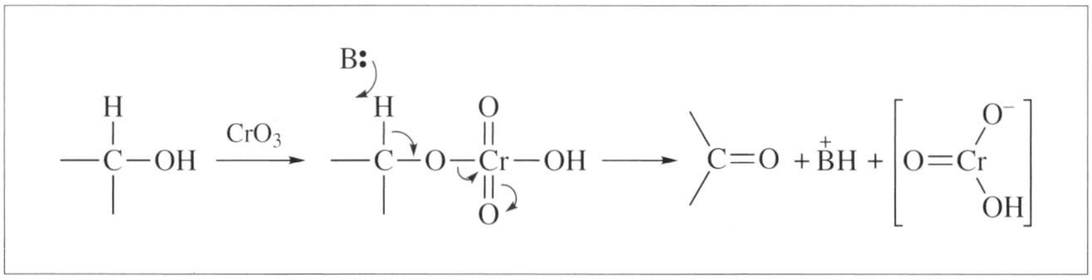

H —C—OH $\xrightarrow{CrO_3}$ B: H —C—O—Cr(=O)(=O)—OH ⟶ >C=O + $\overset{+}{B}H$ + [O=Cr(O⁻)OH]

Scheme 7.2

ethynyl alcohols in propanone solution, but it has since found widespread application.

The oxidation of primary alcohols proceeds through the aldehyde to the carboxylic acid. The second step in this oxidation involves the fragmentation of the chromate(VI) ester of the hydrate of the aldehyde (Scheme 7.3). The hydration of the aldehyde is an acid-catalysed process. The chromium(VI) oxide–pyridine complex ($CrO_3 \cdot C_5H_5N$) was introduced to prevent further oxidation of the aldehyde. The presence of the pyridine diminishes the chances of acid-catalysed side reactions. Other chromium(VI) oxide complexes, such as a chromium(VI) oxide–dimethylpyrazole complex, have also been introduced, since they incorporate an internal base to help decomposition of the intermediate chromate(VI) ester.

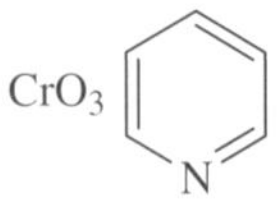

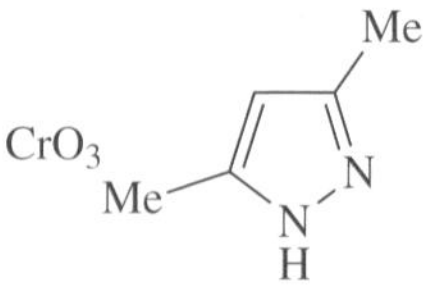

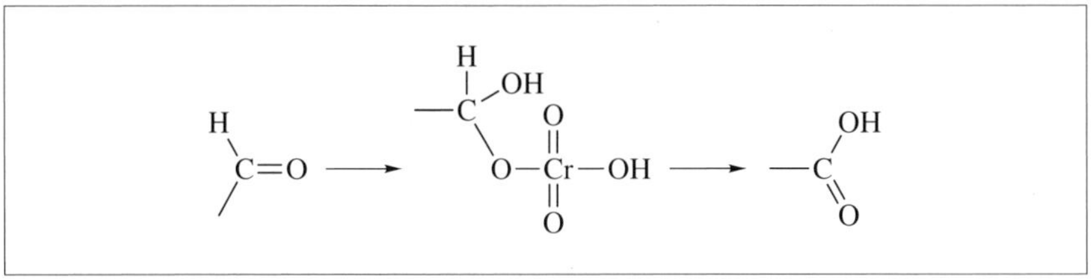

Scheme 7.3

The use of **silver(I) oxide** (Ag_2O) to oxidize aldehydes to carboxylic acids involves the nucleophilic addition of the silver oxide to the aldehyde, followed by the reduction of the silver oxide to silver (the silver mirror test for aldehydes) (Scheme 7.4).

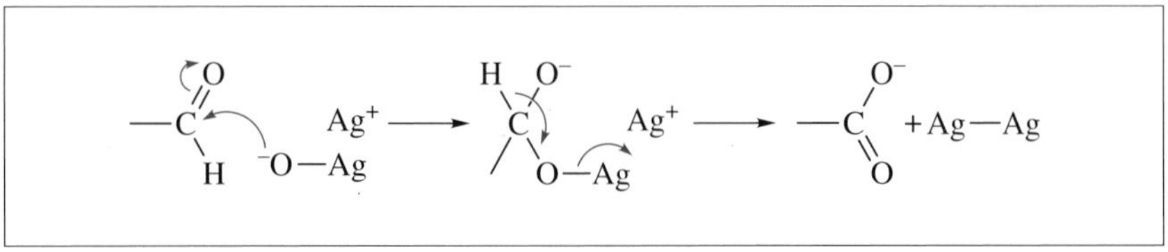

Scheme 7.4

A number of reagents have been devised for the specific oxidation of alcohols to aldehydes, avoiding the over-oxidation to carboxylic acids. **Trimethylamine *N*-oxide** (Me_3NO) is an example. The alcohol is converted to its toluene-*p*-sulfonate, which then undergoes a nucleophilic displacement by trimethylamine *N*-oxide. The resultant trimethylammonium salt (Scheme 7.5) undergoes a base-catalysed fragmentation in the presence of potassium carbonate with the elimination of trimethylamine. There is nothing present which might oxidize the aldehyde to an acid.

$Me_3N{\rightarrow}O$
Trimethylamine *N*-oxide

Dimethyl sulfoxide ($Me_2S{=}O$, DMSO) is another relatively mild oxidant which is used in the **Swern oxidation** of alcohols. In the first step the dimethyl sulfoxide is activated by treatment with oxalyl chloride to form a dimethylchlorosulfonium salt (Scheme 7.6). This reacts with the

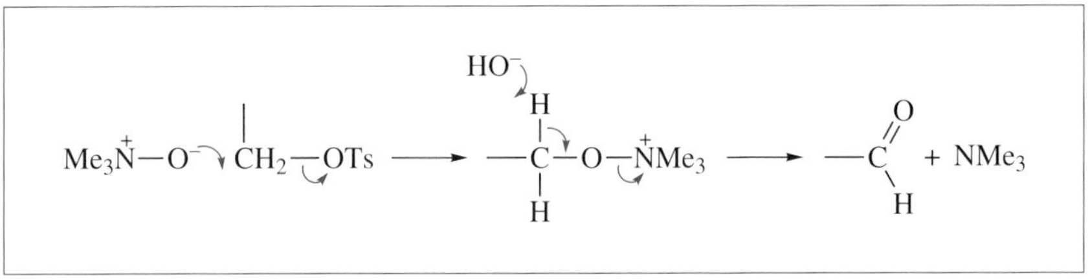

Scheme 7.5

alcohol and the sulfoxonium salt then fragments, with the elimination of dimethyl sulfide.

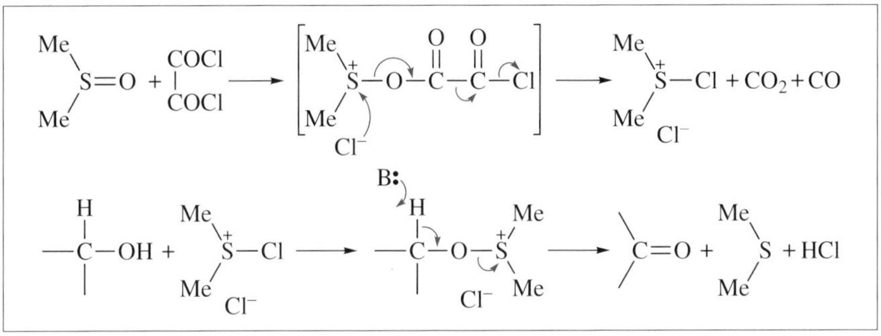

Scheme 7.6 Swern oxidation

The **Dess–Martin periodinane oxidation** is another mild reaction based on the same fragmentation principle. The periodinane (Scheme 7.7) is obtained from 2-iodoxybenzoic acid and acetic anhydride. An alcohol displaces an acetate unit to give an intermediate, which, in turn, undergoes a fragmentation, leading to oxidation of the alcohol.

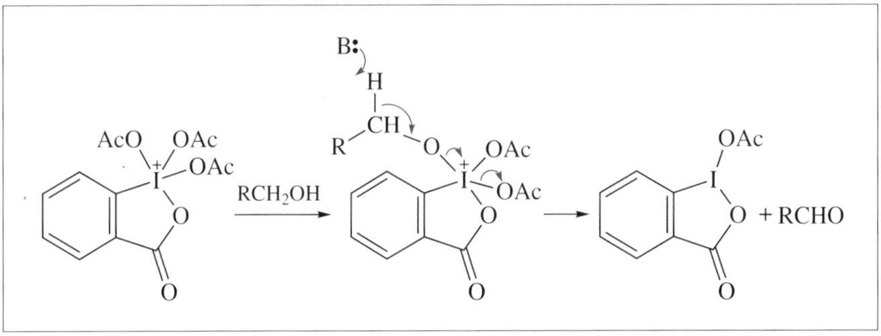

Scheme 7.7 Dess–Martin oxidation

Tetra-*n*-propylammonium perruthenate(VII) ($Pr_4N^+RuO_4^-$, TPAP) is a mild reagent which can be used catalytically, with *N*-methylmorpholine *N*-oxide (NMO) as the oxidant to re-oxidize the catalyst. It can be used in the presence of quite sensitive protecting groups such as the tetrahydropyranyl (THP) and *tert*-butyldimethylsilyl (TBDMS) ethers (Scheme 7.8). Although **potassium permanganate** [potassium manganate(VII)] has been a standard reagent for the oxidation of alcohols, it is less specific and a number of other functional groups, such as alkenes, are also oxidized. Manganese dioxide has been used for the specific oxidation of allylic alcohols to unsaturated ketones.

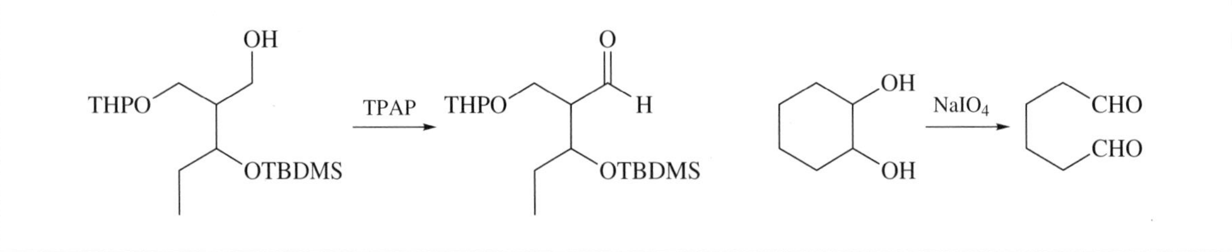

Scheme 7.8

$NaIO_4$
Sodium periodate(VII)

Sodium periodate [sodium iodate(VII)] is a specific oxidant for 1,2-diols, with which it can form a cyclic ester. The breakdown of these cyclic esters leads to dialdehydes (Scheme 7.8). Since potassium permanganate will oxidize alkenes to diols, the combination of these two reagents provides a useful way of breaking carbon–carbon bonds.

Worked Problem 7.1

Q Propose an oxidant for the conversion of **1** to **2**.

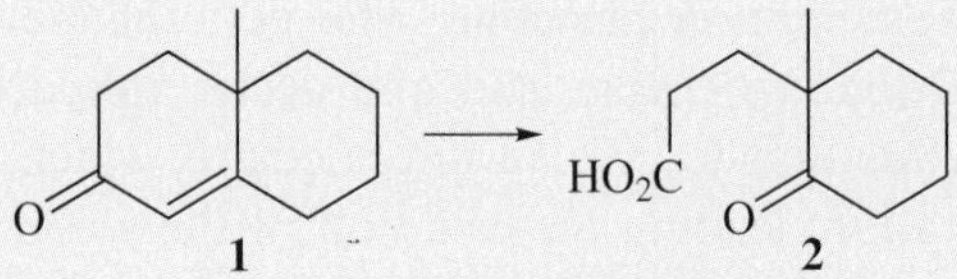

A The oxidation of **1** involves breaking a C=C bond and eliminating one carbon atom. Potassium permanganate will convert an alkene to a diol and sodium periodate will cleave diols and α-hydroxy ketones:

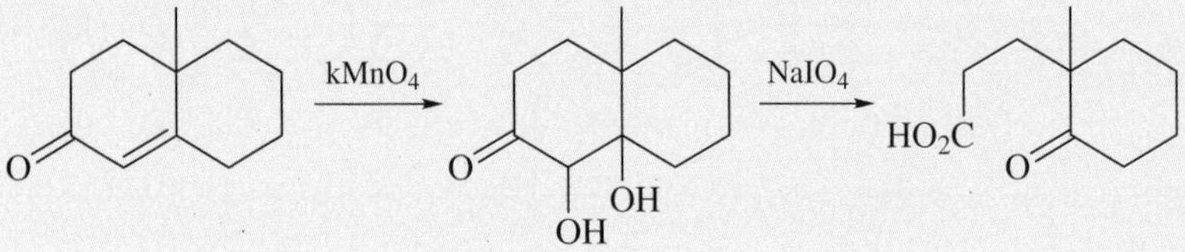

7.1.3 One-electron Oxidation

When there is a one-electron redox change in a molecule, the product is a free radical. In many oxidative reactions, these radicals may dimerize. The best known of these reactions is phenol coupling (see Chapter 5)

7.1.4 Oxidation of Alkenes

The double bond is an electron-rich part of a molecule and it undergoes a number of oxidative addition reactions. The reaction with **ozone** leads first to the molozonide and then to the ozonide. Decomposition of the ozonide by treatment with zinc or triphenylphosphine results in the cleavage of the alkene. Since an alkene may be reconstructed using a Wittig reaction, this makes a useful procedure for labelling a molecule using a ^{14}C-labelled Wittig reagent. Thus, the hydrocarbon ent-kaurene (Scheme 7.9) was shown to be a precursor of the plant hormone gibberellic acid by labelling it with carbon-14 via an ozonolysis and Wittig sequence.

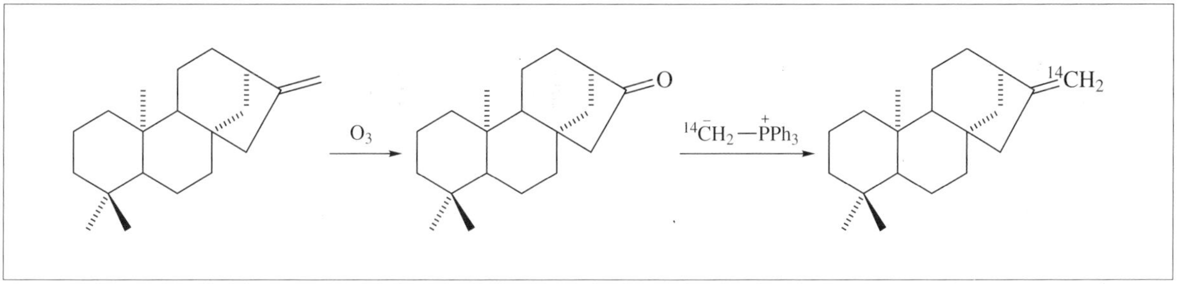

O_3; $^{14}\bar{C}H_2-\overset{+}{P}Ph_3$; $^{14}CH_2$; O

Scheme 7.9

The epoxidation of an alkene with a peroxy acid such as ***m*-chloroperbenzoic acid** (MCPBA) involves a nucleophilic attack of the π-electron cloud of the alkene on the peroxy acid with the formation of the epoxide (Scheme 7.10).

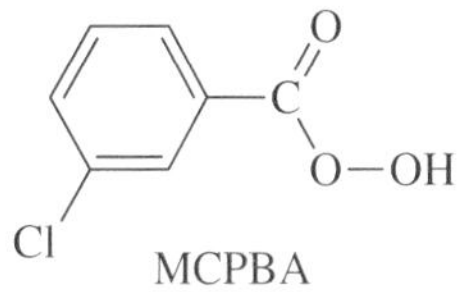

O; C; O—OH; Cl; MCPBA

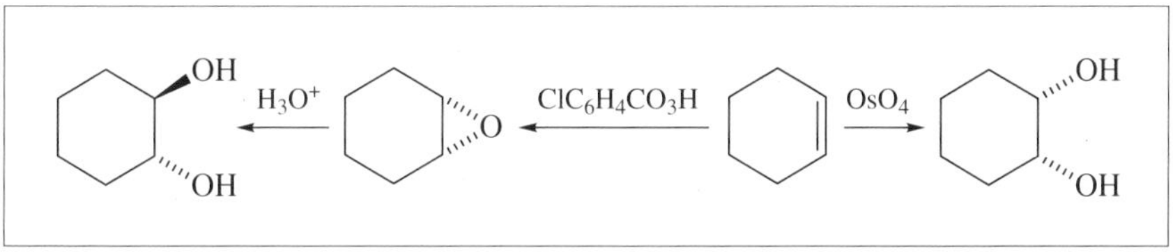

OH; OH; H_3O^+; O; $ClC_6H_4CO_3H$; OsO_4; OH; OH

Scheme 7.10

Hydrolysis of a cyclic epoxide gives a *trans*-1,2-diol which, in a rigid system, will have the diaxial conformation. *cis*-1,2-Diols may be obtained from alkenes by reaction with **osmium(VIII) tetroxide**, either on its own (Scheme 10) or by using it catalytically with a co-oxidant such as *t*-butyl hydroperoxide or potassium hexacyanoferrate(III). Osmylation and epoxidation take place from the less-hindered face of a molecule. Epoxidation using titanium tetraisopropoxide, *t*-butyl hydroperoxide and (+)- or (–)-diethyl tartrate (the **Sharpless reagent**) affords chiral epoxides and hence a range of chiral compounds.

OsO_4
Osmium tetroxide

The use of chiral ligands together with catalytic osmylation leads to the asymmetric dihydroxylation of alkenes. High enantioselectivity has been achieved with ligands derived from the alkaloid quinine.

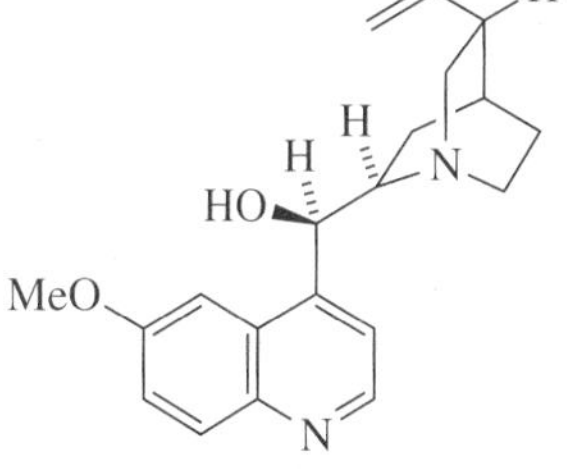

H; H; H; N; HO; MeO; N

quinine

The oxidation of allylic and benzylic positions can be useful in synthesis. The electron-deficient chromium of chromium(VI) oxide or chromyl(VI) chloride reacts with an alkene to form a complex from which the allylic position may be attacked (Scheme 7.11).

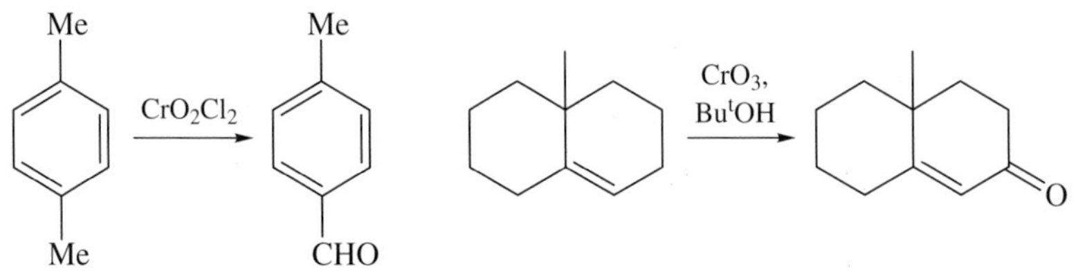

Me, CrO_2Cl_2, Me, Me, CHO; CrO_3, Bu^tOH, O

Scheme 7.11

Worked Problem 7.2

Q Propose reagents for the preparation of the diols **4** and **5** from the alkene **3**.

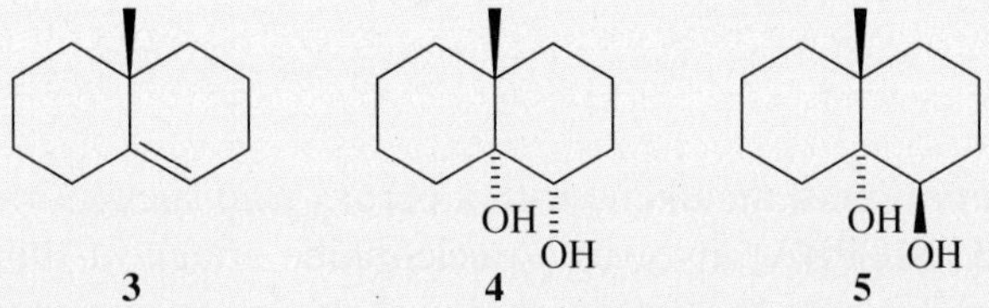

OH OH OH OH
3 4 5

A The *cis* diol may be prepared using osmium(VIII) tetroxide with a co-oxidant (*e.g.* *t*-butyl hydroperoxide). The presence of the methyl group directs the addition to the opposite face of the alkene. The *trans* diaxial diol may be prepared by the diaxial hydrolysis of the epoxide. The initial epoxidation of the alkene by a peroxy acid is directed to the opposite face by the methyl group:

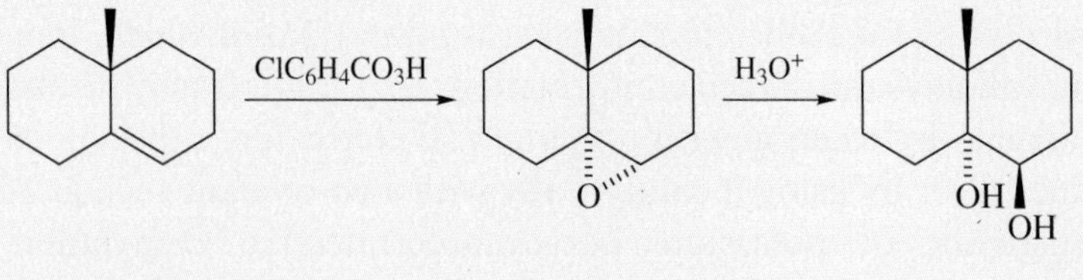

$ClC_6H_4CO_3H$, O, H_3O^+, OH, OH

7.1.5 The Baeyer–Villiger Oxidation

The treatment of ketones with a peroxy acid in the presence of an acid catalyst (**Baeyer–Villiger oxidation**) leads to the conversion of a ketone to an ester or, in the case of a cyclic ketone, to a lactone. This can provide not only a useful method of opening a carbocyclic ring, but also of introducing functionality into a chain (Scheme 7.12).

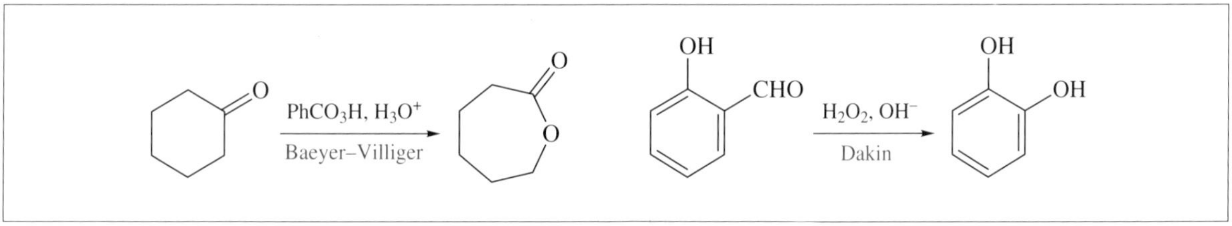

$$\text{cyclohexanone} \xrightarrow[\text{Baeyer–Villiger}]{PhCO_3H,\ H_3O^+} \text{caprolactone} \qquad \text{salicylaldehyde (OH, CHO)} \xrightarrow[\text{Dakin}]{H_2O_2,\ OH^-} \text{catechol (OH, OH)}$$

Scheme 7.12 Baeyer–Villiger and Dakin oxidations

A reaction which has a similar outcome is the **Dakin oxidation** of aromatic ketones, which leads to phenols. Since it is easier to acylate a ring rather than introduce a hydroxyl group directly, this provides a useful method of preparing polyhydroxylic phenols (Scheme 7.12).

7.2 Reduction

Reductive processes can be divided into three groups: (1) hydrogenolysis reactions involving the substitution of a group by hydrogen; (2) hydrogenation reactions involving the addition of hydrogen to an unsaturated system; and (3) reactions involving the removal of oxygen without necessarily including the addition of hydrogen. The reagents that are used may be divided into four classes. These are: (i) catalytic systems involving hydrogen gas and a catalyst; (ii) "dissolving metal" systems which involve the donation of electrons to a substrate from a metal reacting either with an acid or with an alcohol; (iii) hydride reagents involving the nucleophilic attack of a hydride ion; and (iv) hydrogen transfer reagents.

7.2.1 Hydrogenolysis Reactions

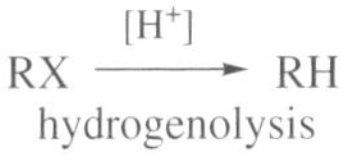

Hydrogenolysis reactions involve the replacement of a substituent by hydrogen. The key step in these reactions may involve either the nucleophilic substitution of a good leaving group by a hydride, or the homolytic fission of a bond between a carbon atom and a substituent and the removal of a hydrogen atom by this radical or the protonation of a carbanion derived from an organometallic reagent.

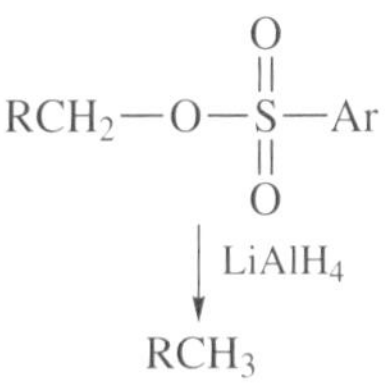

$$RX \xrightarrow[\text{hydrogenolysis}]{[H^+]} RH$$

The nucleophilic substitution of a toluene-*p*-sulfonate or methanesulfonate of an alcohol, by a hydride reagent such as **lithium aluminium hydride** ($LiAlH_4$), proceeds with inversion of configuration. It is a useful way of converting an alcohol to a hydrocarbon. Since hydride reagents are available in a labelled form, for example lithium aluminium deuteride, this has been employed in labelling compounds.

$$RCH_2{-}O{-}S(=O)_2{-}Ar \xrightarrow{LiAlH_4} RCH_3$$

Tri-*n*-butyltin hydride (Bu_3SnH) has been widely used for the radical dehalogenation of organic compounds, particularly bromides and iodides. As a radical reagent, it requires an initiator, typically azobisisobutyronitrile (AIBN). The tin radical abstracts a halogen atom from the substrate to form a carbon radical, which then participates in a radical chain process removing a hydrogen atom from another molecule of tri-*n*-butyltin hydride.

Other hydrogenolysis reactions which possibly have a radical character involve the use of **sodium in ammonia** and **Raney nickel** (see Section 7.2.3) for the cleavage of a carbon–sulfur bond. The hydrogenolysis of thioketals, which may be obtained from ketones, provides a mild method for converting a ketone into a methylene group.

$$R_2C(SR)_2 \xrightarrow{H_2,\ Ni} R_2CH_2$$

Organometallic reagents such as **Grignard reagents** are obtained from alkyl halides. Since metals are more electropositive than carbon, these are sources of carbanions. Protonation of these organometallic compounds leads to the overall replacement of halogen by hydrogen. This can be a method for specifically labelling compounds, since the acid concerned may be derived from ^2H_2O.

$$RMgCl + {}^2H_2O \longrightarrow R^2H$$

Hydrogenolysis reactions are quite widely used in removing protecting groups. The direction of fission of a benzyl– or triphenylmethyl–oxygen bond is governed by the stabilizing effect of the aromatic ring on a benzyl or triphenylmethyl cation or radical. These hydrogenolysis reactions leave the bond between the heteroatom and the substrate intact. They are widely used in peptide synthesis.

7.2.2 Hydrogenation Reactions

The major groups of unsaturated compound that undergo reduction reactions are alkenes, alkynes and arenes with symmetrical π-bonds, and carbonyl compounds, nitriles, imines and nitro compounds with unsymmetrical π-bonds. Table 7.1 shows the broad selectivity of the various types of reagent for these substrates.

Table 7.1 General patterns of reduction[a]

Substrate	*Reagent*			
	Catalytic (H_2, Pd or Pt)	*Dissolving metal (Na, liq. NH_3)*	*Hydride ($LiAlH_4$)*	*Hydride transfer (N_2H_4)*
C=C	✓	×	×	Dienes
C≡C	✓	✓	Some	×
Ar	✓ (pressure)	✓ (activated)	×	×
C=O	✓	✓	✓	✓
NO_2	✓	✓	✓	Some

[a] ✓ = reacts; × = unreactive.

7.2.3 Catalytic Reductions

Catalytic reduction may involve either heterogeneous or homogeneous catalysis. The heterogeneous catalysts, such as **platinum** or **palladium** absorbed on charcoal or barium sulfate, or a finely divided form of nickel known as **Raney nickel**, function by adsorbing both hydrogen and the substrate onto the catalyst surface. This leads to the *cis* transfer of hydrogen to the substrate. Steric factors play an important role in determining the orientation of the substrate on the surface and hence the face of the alkene to which the hydrogen is added. Thus the hydrogenation of α-pinene gives the dihydro compound (Scheme 7.13). However, this is not the only feature, and the stereochemistry of hydrogenation can be sensitive to the pH of the medium. Furthermore, double bond isomerization may occur while hydrogen exchange can happen at allylic positions. The extent (Scheme 7.14) of the catalytic tritiation of *trans*-but-2-en-1-ol at the various positions reflects these factors.

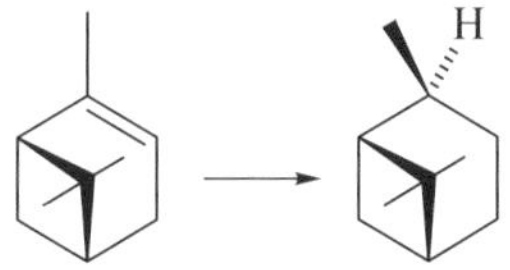

Scheme 7.13

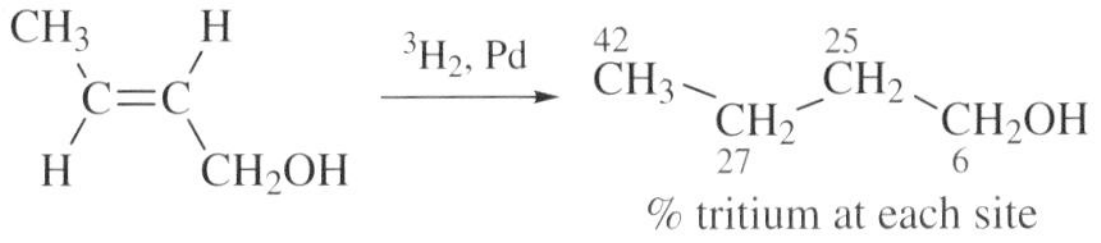

$CH_3CH{=}CHCH_2OH \xrightarrow{^3H_2,\ Pd} CH_3\text{–}CH_2\text{–}CH_2\text{–}CH_2OH$ (42, 27, 25, 6)

% tritium at each site

Scheme 7.14

It is possible to modify catalysts so that they are suitable for specific purposes. For example, a palladium catalyst supported on barium sulfate and partially poisoned with quinoline (the **Lindlar catalyst**) may be used for the hydrogenation of an alkyne to a *cis*-alkene. A poisoned catalyst may be used for the reduction of an imino chloride to an aldehyde (the **Rosenmund reduction**).

A number of metal salts form alkene complexes. Since these metals also form hydrides, there is the potential for a reaction between the alkene and the hydrogen ligand. This has led to the development of a family of soluble homogeneous catalysts for hydrogenation, of which the best known is **Wilkinson's catalyst** [$(Ph_3P)_3RhCl$].

These complexes have various amines and phosphines as ligands, and these can play a part in the catalytic process. If the ligands are chiral, it is possible to induce chirality in the product of the hydrogenation. An

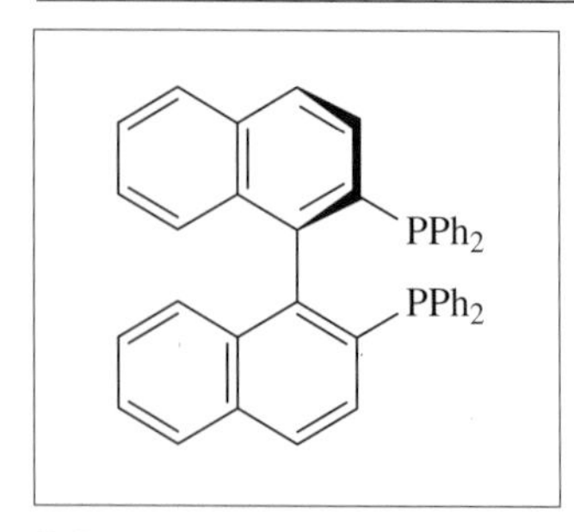

Scheme 7.15

example of these ligands is the binaphthylphosphine **BINAP** (Scheme 7.15), which is chiral as a result of hindered rotation about the bond between the two naphthalene rings.

7.2.4 Dissolving Metal Reductions

Electropositive metals, such as **sodium** dissolving in ethanol or ammonia or **zinc** dissolving in hydrochloric acid, are powerful reductants. The sodium in ethanol reduction of esters to alcohols is known as the **Bouveault–Blanc reduction**. The reduction of ketones proceeds by the stepwise delivery of electrons to the system:

$$R_2C{=}O + Na \longrightarrow R_2\dot{C}{-}O^-Na^+ \xrightarrow{Na} R_2\bar{C}{-}O^-\,2Na^+$$

Since the reaction proceeds via a carbanion, this is free to take up the most stable conformation before the final protonation which completes the reduction. The stereochemistry of the product is determined by the relative stability of the different epimers.

Reductions which use sodium or lithium dissolving in liquid ammonia are known as the **Birch reduction**. This reduction is often applied to the reduction of methoxybenzenes. Protonation of the radical anions leads to enol ethers, which on subsequent hydrolysis give α,β-unsaturated ketones (Scheme 7.16).

OMe → (Na, NH_3, Bu^tOH) → OMe → (H_3O^+) → O

Scheme 7.16 Birch reduction

The value of this in synthesis is that a carbon skeleton can be constructed using aromatic chemistry, with electrophilic methods of making carbon–carbon bonds and minimizing stereochemical problems. Conversion to the unsaturated ketone changes the reactive character of the carbon atoms and allows for thermodynamic control of the stereochemistry of any ring junctions that are involved.

$$Me_2C{=}O \longrightarrow Me_2C(OH){-}C(OH)Me_2$$

The dissolving metal reduction of ketones may lead to dimeric products (pinacols). This is particularly the case when the reductant is a metal such as **magnesium,** which coordinates with oxygen functions. **Titanium** forms sufficiently strong bonds to oxygen to remove the oxygen from the dimeric species as titanium dioxide, with the creation of an alkene rather than a diol (the **McMurry reaction**) (see Chapter 5).

When **zinc** and concentrated hydrochloric acid is the reductant, the final outcome from the reduction of a ketone is a methylene group in the **Clemmensen reduction**

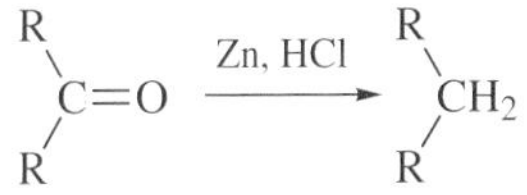

$$R_2C{=}O \xrightarrow{\text{Zn, HCl}} R_2CH_2$$

The dissolving metal reduction of aromatic nitro compounds provides a standard method for the preparation of aromatic amines. The reactions are normally carried out with **iron** or **tin** in hydrochloric acid. These form salts with the amine and prevent further reaction. If, however, the reduction is carried out under milder conditions (*e.g.* Zn and NaOH), the amine may react with unchanged nitro compound to give dimeric products such as azoxy and azo compounds (Scheme 7.17).

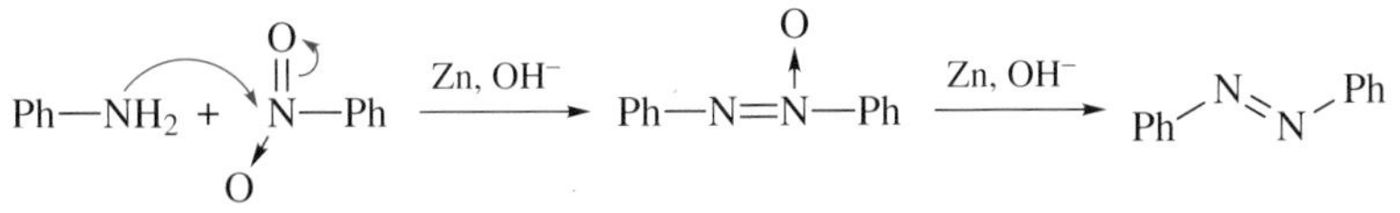

$$Ph{-}NH_2 + Ph{-}NO_2 \xrightarrow{\text{Zn, OH}^-} Ph{-}N{=}N(\rightarrow O){-}Ph \xrightarrow{\text{Zn, OH}^-} Ph{-}N{=}N{-}Ph$$

Scheme 7.17

The dissolving metal reduction of other functional groups containing nitrogen, such as oximes and nitriles, also provides routes to amines.

7.2.5 Hydride Reagents

A series of hydride reagents is available with a reactivity that ranges from the vigorous **lithium aluminium hydride** to milder reagents such as **lithium tri(*t*-butoxy)aluminium hydride** and **sodium borohydride**. The selectivity of these hydrides against a number of the common functional groups is given in Table 7.2. This table is a general guide. The substrate stereochemistry, the solvent and the presence of additional metal salts can all have a significant effect on the outcome.

The hydride reagents are a source of nucleophilic hydrogen. The major determining feature in the stereochemistry of the reaction is the path of the reagent as it approaches the carbonyl group. This may be affected both by the bulk of the reagent and the presence of metal salts which coordinate to the oxygen of a carbonyl group. Metal salts affect the

Table 7.2 Reductions with hydride reagents[a]

Substrate	*$LiAlH_4$*	*$NaBH_4$*	*$LiBH_4$*	*$LiAlH(OBu^t)_3$*	*B_2H_6*
RCHO	✓	✓	✓	✓	✓
RCOR	✓	✓	✓	✓	✓
RCO_2H	✓	×	×	×	✓
RCO_2R	✓	×	✓	×	Slow
RCOCl	✓	✓	✓	×	Slow

[a] ✓ = reacts; × = unreactive.

regiochemistry of the reduction. For example, α,β-unsaturated ketones may be reduced to the saturated ketone and alcohol by sodium borohydride in the presence of copper salts, whereas the presence of cerium salts leads to the formation of allylic alcohols.

$$\text{>CH-CH(-CHOH-)} \xleftarrow{\text{NaBH}_4\ |\ \text{CuCl}} \text{>C=C(-C=O)} \xrightarrow{\text{NaBH}_4\ |\ \text{CeCl}_3} \text{>C=C(-CHOH-)}$$

Worked Problem 7.3

Q Propose a reagent for the conversion of **6** to **7**.

$$\underset{\mathbf{6}}{\text{MeCH=CHCH=O}} \longrightarrow \underset{\mathbf{7}}{\text{MeCH=CHCH}_2\text{OH}}$$

A There are two functional groups in **6** that are sensitive to reduction. A hydride reagent would reduce the carbonyl group of the aldehyde. To avoid 1,4-addition, the reduction might be carried out with sodium borohydride in the presence of cerium(III) chloride:

$$\text{MeCH=CHCH=O} \xrightarrow[\text{CeCl}_3]{\text{NaBH}_4} \text{MeCH=CHCH}_2\text{OH}$$

Worked Problem 7.4

Q Propose a method for the selective reduction of 3,5-dinitrobenzoic acid (**8**) to 3,5-dinitrobenzaldehyde (**9**).

$$\underset{\mathbf{8}}{3,5\text{-}(O_2N)_2C_6H_3CO_2H} \longrightarrow \underset{\mathbf{9}}{3,5\text{-}(O_2N)_2C_6H_3CHO}$$

A The reagent must differentiate between the carboxylic acid or a derivative and the nitro group. A suitable procedure is to convert the carboxylic acid to the reactive acyl chloride and then use a mild version of one of the hydride reagents such as lithium tri(*t*-butoxy)aluminium hydride at a low temperature:

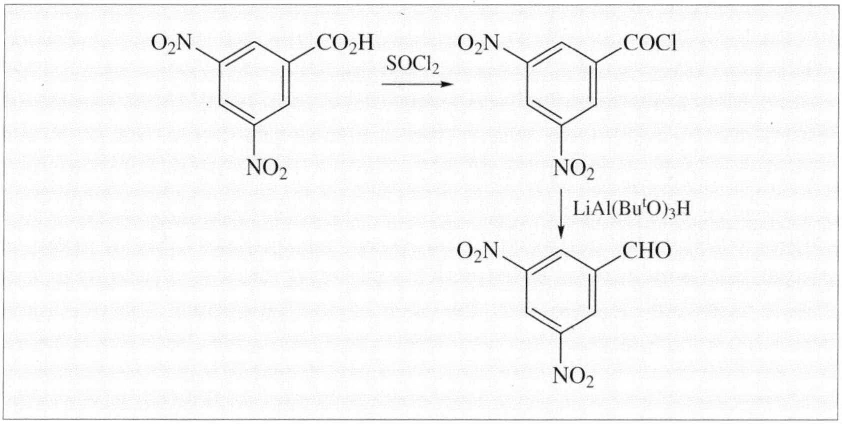

7.2.6 Hydride Transfer Reagents

The **Wolff–Kishner reduction** of a ketone to a methylene group, by the decomposition of a hydrazone by alkali, is a vigorous but effective procedure. The reaction involves the isomerization of the hydrazone to an azine, which undergoes decomposition (Scheme 7.18). The Huang-Minlon variation involves heating the hydrazone with alkali to 200 °C, although the use of potassium *t*-butoxide in dimethyl sulfoxide permits milder conditions.

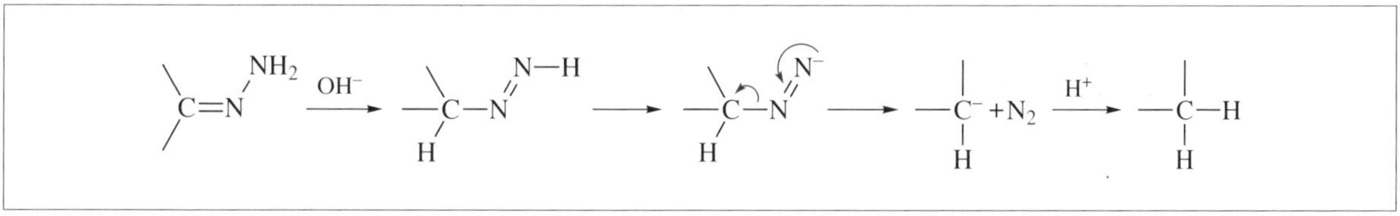

Scheme 7.18 Wolff–Kischner reduction

When the reaction is carried out using a toluene-*p*-sulfonylhydrazone, the product is an alkene (Scheme 7.19). If the substrate is an α,β-epoxy ketone, fragmentation occurs in the Wolff–Kishner reduction with

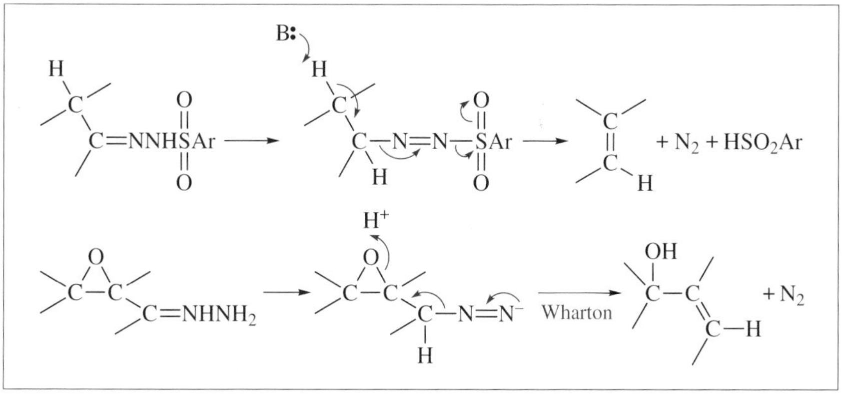

Scheme 7.19 Hydride transfer, including the Wharton reaction

the formation of an allylic alcohol (the **Wharton reaction**) (Scheme 7.19).

The **Meerwein–Ponndorf reduction** of ketones is another example of a hydride transfer reduction. Aluminium isopropoxide mediates the transfer of hydrogen from an alcohol such as propan-2-ol (isopropanol) to a ketone through a six-membered transition state (Scheme 7.20). The propan-2-ol, which is present in excess, is oxidized in the process to propanone, which may be distilled out of the reaction mixture. The procedure is reversible, and may be used as an oxidation if the hydrogen acceptor is in excess.

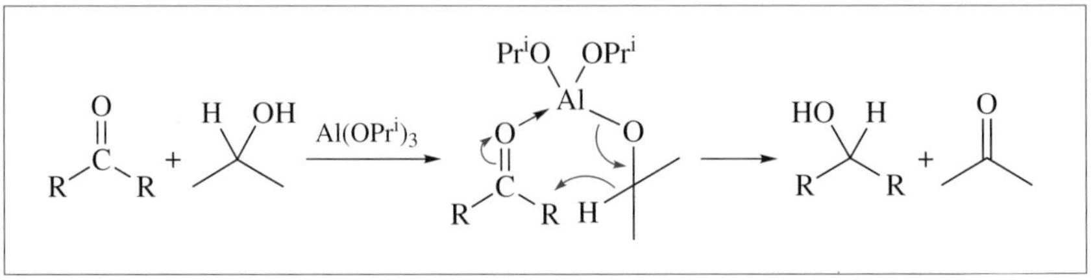

Scheme 7.20
Meerwein–Ponndorf reduction

7.2.7 Hydroboration

Borane adds in a *cis* manner to an alkene to give an alkylborane. The transformations of these alkylboranes are of major synthetic importance. The monomeric BH_3 is stabilized as its tetrahydrofuran complex, or as a complex with dimethyl sulfide or trimethylamine. It is this monomeric form which adds to alkenes. The electron-deficient borane first forms a π-complex with the alkene (Scheme 7.21). The facial selectivity of the addition to the alkene is determined by the steric hindrance of neighbouring alkyl groups to the approaching borane, and by the influence of the lone pairs on adjacent oxygen functions on the electron density of the different faces of the alkene. The second stage in the reaction involves the rearrangement of the π-complex to form an alkylborane. The regiochemistry of this is determined by the stabilization of carbocations at the respective carbon atoms. In Markownikoff addition the electrophile (*e.g.* H^+) becomes attached to the carbon atom bearing the most hydrogen atoms. The nucleophile (*e.g.* OH^-) subsequently adds to the more highly substituted carbon atom. The electron-deficient borane behaves in the same way as a proton, resulting in the addition of the borane to

Scheme 7.21

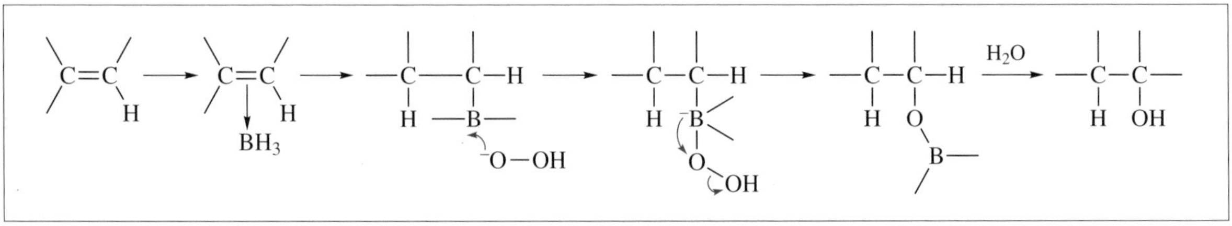

the carbon atom bearing the more hydrogen atoms. The hydrogen from the borane behaves as though it was a nucleophile (Scheme 7.21).

The commonest transformation of the alkylborane involves its decomposition with alkaline hydrogen peroxide. The highly nucleophilic hydroperoxide anion attacks the electron-deficient boron with the formation of an ate complex (Scheme 7.21). Rearrangement of this leads to the formation of a borate ester, which then undergoes hydrolysis to an alcohol in which an oxygen atom has replaced the boron (the original electrophile). The result of the reaction is the equivalent of the **anti-Markownikoff hydration** of the double bond. The overall addition of water occurs in a *cis* manner on the less-hindered face of the alkene (Scheme 7.22).

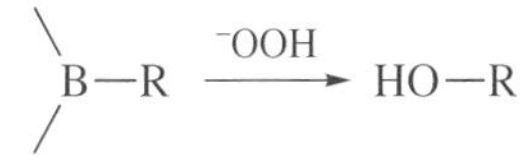

$>B{-}R \xrightarrow{^{-}OOH} HO{-}R$

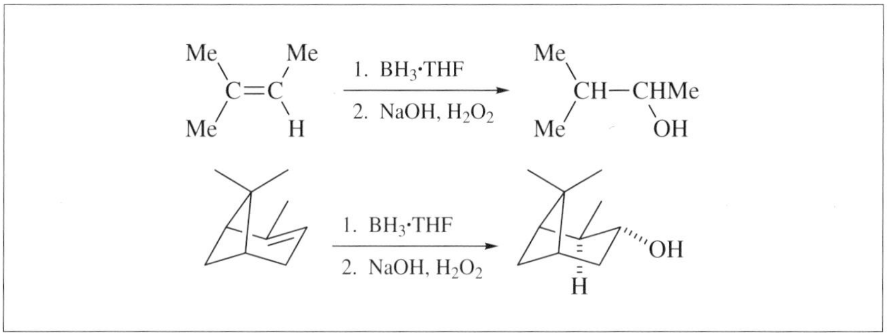

$Me_2C{=}C(Me)H \xrightarrow[2.\ NaOH,\ H_2O_2]{1.\ BH_3{\cdot}THF} Me_2CH{-}CH(OH)Me$

1. $BH_3{\cdot}THF$ 2. NaOH, H_2O_2 → OH, H

Scheme 7.22

Another useful reaction of boranes involves protonolysis of the borane with a carboxylic acid. This leads to hydrogenation of the alkene.

Oxidation of a borane with chromium(VI) oxide may lead directly to a ketone. Boranes may be replaced with halogens using bromine or iodine monochloride for iodinations. Aminations with hydroxylamine-*O*-sulfonic acid have also been reported.

Carbonylation reactions of alkylboranes can be used to synthesize aldehydes, ketones and tertiary alcohols and have been described in Chapter 2.

In order to enhance the regio- and stereoselectivity of these reactions, the more highly substituted boranes such as **thexylborane** ("*t*-hexyl") and **9-BBN** (9-borabicyclononane) may be used in the initial addition to the alkene. They are prepared from 2,3-dimethylbut-2-ene and cycloocta-1,5-diene, respectively (Scheme 7.23).

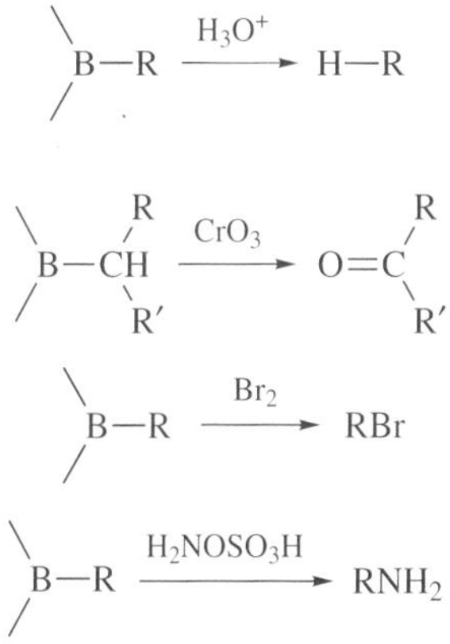

$>B{-}R \xrightarrow{H_3O^+} H{-}R$

$>B{-}CHRR' \xrightarrow{CrO_3} O{=}CRR'$

$>B{-}R \xrightarrow{Br_2} RBr$

$>B{-}R \xrightarrow{H_2NOSO_3H} RNH_2$

7.3 Halogenation

The introduction of halogen into a molecule can be an important part of a synthetic sequence as a prelude to the formation of a new carbon–carbon bond via an organometallic method or a coupling reaction. Furthermore, a number of biologically active compounds contain one or more halogen atoms.

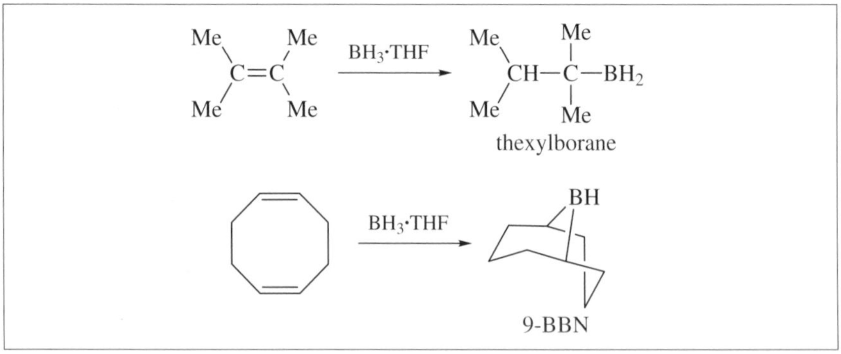

$Me_2C{=}CMe_2 \xrightarrow{BH_3 \cdot THF} Me_2CH{-}CMe_2{-}BH_2$

thexylborane

cyclooctadiene $\xrightarrow{BH_3 \cdot THF}$ 9-BBN

Scheme 7.23

The halogen can be introduced into a molecule by nucleophilic, electrophilic and radical halogenations.

7.3.1 Nucleophilic Substitution of Alcohols

Methods for the conversion of alcohols into alkyl halides involve weakening and then breaking the carbon–oxygen bond in the sense $C^{\delta+}$–$O^{\delta-}$. The reaction of an alcohol with concentrated hydrochloric acid and zinc chloride, or with concentrated hydrobromic acid in sulfuric acid, takes place under vigorous conditions in which a mineral or Lewis acid attacks the lone pairs on the oxygen, helping the cleavage of the carbon–oxygen bond. The halide may then attack the carbon atom. However, the conditions are sufficiently vigorous for side reactions, including molecular rearrangements, to take place.

A number of phosphorus and sulfur halides and oxyhalides (Scheme 7.24) bring about nucleophilic substitution of an alcohol by first forming an ester, such as a chlorosulfite. This then acts as a leaving group to facilitate the nucleophilic substitution. The reaction is carried out in the presence of a base to remove the protons that are released. Whereas **thionyl chloride** ($SOCl_2$) tends to give chlorides with retention of configuration, **phosphorus pentachloride** gives inversion of configuration.

Many methods involve separating the activation step from the substitution step by first converting the alcohol into a good leaving group and then carrying out the substitution. Thus an alcohol may be converted to its toluene-*p*-sulfonate and this is then treated with an inorganic halide (lithium bromide or sodium iodide) in a solvent of relatively high dielectric constant (such as dimethylformamide, DMF) to promote ionization.

Activation may be carried out with reagents based on elements that form strong bonds to oxygen. The trimethylsilyl derivatives of alcohols can be displaced by bromides or iodides, using sodium bromide or iodide in a solvent such as DMF or acetonitrile.

Phosphorus forms strong bonds to oxygen. Consequently, there are a

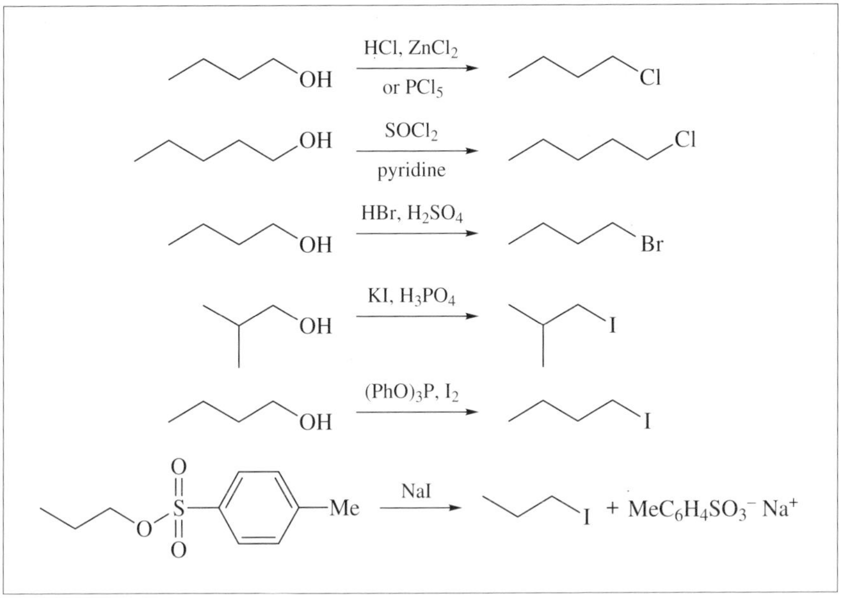

Scheme 7.24

number of methods for converting alcohols to alkyl halides that are based on phosphorus activation. **Triphenyl phosphite methiodide** [methyl-(triphenoxy)phosphonium iodide], prepared from triphenyl phosphite and iodomethane, reacts with an alcohol with the displacement of phenol. The phosphorus derivative is a good leaving group and forms a phosphonate. Hence attack by an iodide takes place with inversion of configuration (Scheme 7.25).

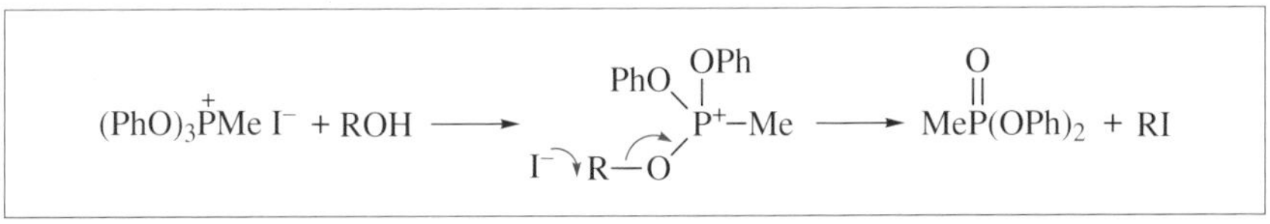

Scheme 7.25

Triphenylphosphine reacts with **tetrachloromethane** or **tetrabromomethane** to form phosphonium salts, which can in turn participate in the activation of alcohols (Scheme 7.26). The reaction normally proceeds with inversion of configuration.

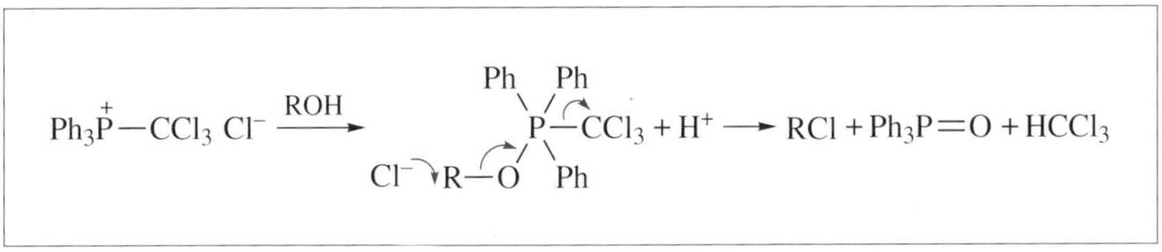

Scheme 7.26

The **Mitsunobu procedure**, which is often used for the inversion of configuration of alcohols, may also be used to convert an alcohol to an alkyl halides (Scheme 7.27). The triphenylphosphine is first activated by reaction with diethyl azodicarboxylate. Reaction of the resultant phosphonium salt with the alcohol gives a reactive intermediate in which the phosphorus–oxygen bond is readily displaced by a nucleophile with inversion of configuration. The reaction of the bis(ethoxycarbonyl)hydrazine anion with an alkyl halide or zinc halide releases a halide ion which provides a nucleophile for displacement of the triphenylphosphine oxide. The simple alkyl halides such as iodomethane, bromomethane or dichloromethane act as a nucleophile carrier in this form of the Mitsunobu reaction. In several steps of this sequence, the ester grouping of the diethyl azodicarboxylate serves to stabilize an adjacent anion.

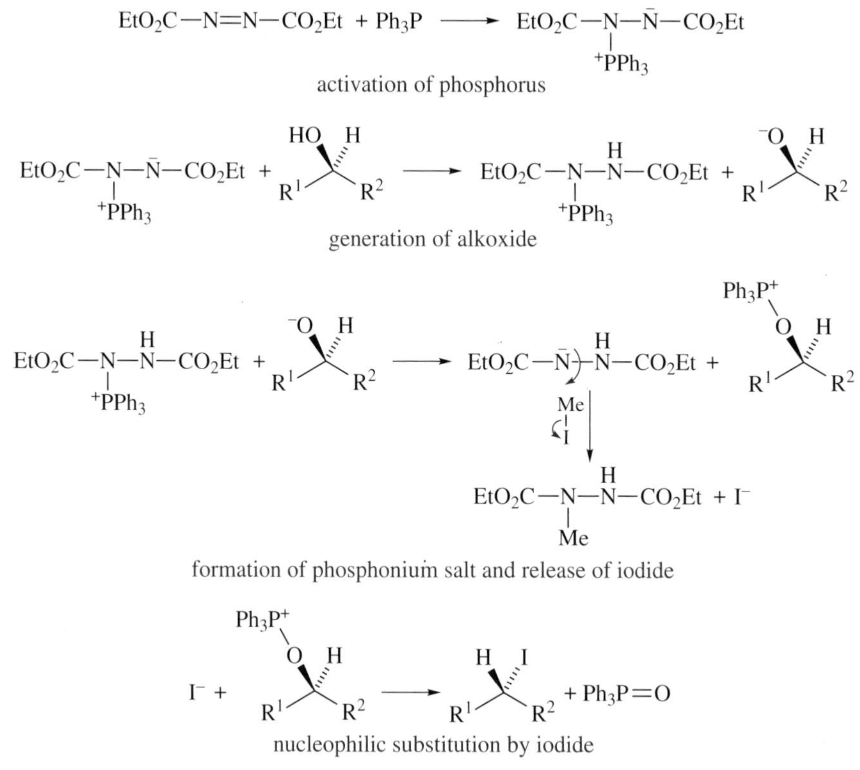

$EtO_2C{-}N{=}N{-}CO_2Et + Ph_3P \longrightarrow EtO_2C{-}N(^+PPh_3){-}\bar{N}{-}CO_2Et$

activation of phosphorus

$EtO_2C{-}N(^+PPh_3){-}\bar{N}{-}CO_2Et$ + HO H R¹ R² ⟶ $EtO_2C{-}N(^+PPh_3){-}N(H){-}CO_2Et$ + ⁻O H R¹ R²

generation of alkoxide

$EtO_2C{-}N(^+PPh_3){-}N(H){-}CO_2Et$ + ⁻O H R¹ R² ⟶ $EtO_2C{-}\bar{N}{-}N(H){-}CO_2Et$ + Ph_3P^+ O H R¹ R²

Me–I ↓

$EtO_2C{-}N(Me){-}N(H){-}CO_2Et + I^-$

formation of phosphonium salt and release of iodide

I^- + Ph_3P^+ O H R¹ R² ⟶ H I R¹ R² + $Ph_3P{=}O$

nucleophilic substitution by iodide

Scheme 7.27 The Mitsunobu reaction

Worked Problem 7.5

Q Propose a method for converting α-hydroxyphenylacetic acid (mandelic acid, **10**) into α-chlorophenylacetic acid (**11**).

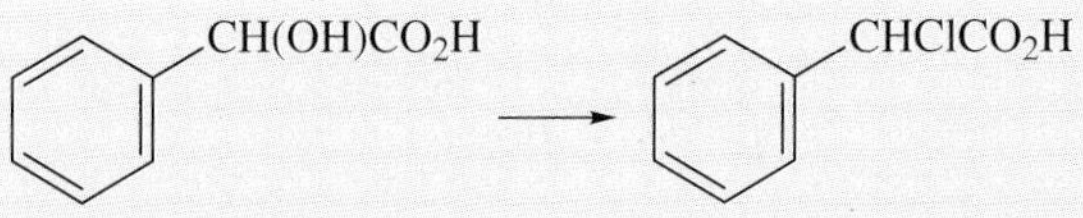

$C_6H_5CH(OH)CO_2H$ (10) → $C_6H_5CHClCO_2H$ (11)

A Although at first sight this may seem to be a simple substitution reaction using thionyl chloride or phosphorus pentachloride, these conditions would also convert the carboxylic acid into the acyl chloride. The carboxylic acid hydroxyl group needs to be protected as an ester prior to the halogenation. A suitable ester might be the ethyl ester or *t*-butyl ester, which can be removed under acidic conditions that do not bring about the nucleophilic substitution. The actual procedure uses an acid-catalysed transesterification:

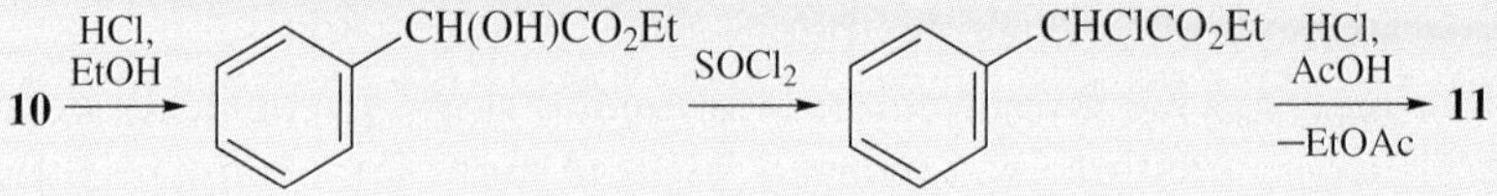

10 —(HCl, EtOH)→ $C_6H_5CH(OH)CO_2Et$ —($SOCl_2$)→ $C_6H_5CHClCO_2Et$ —(HCl, AcOH, −EtOAc)→ 11

7.3.2 Electrophilic Methods

The addition of halogens and hydrogen halides to alkenes to form vicinal dihalides and alkyl halides has been described in the companion text on functional groups. The addition of the halogens to cyclohexenes involves a ***trans*** **diaxial addition** (Scheme 7.28). With trisubstituted alkenes, the addition of the hydrogen halides obeys the Markownikoff rule. However, hydrogen bromide in the presence of peroxides undergoes homolytic fission, and the reaction involves the addition of bromine atoms to the most accessible centre. The best example of the so-called peroxide effect is with allyl bromide (see *Functional Group Chemistry* in Further Reading).

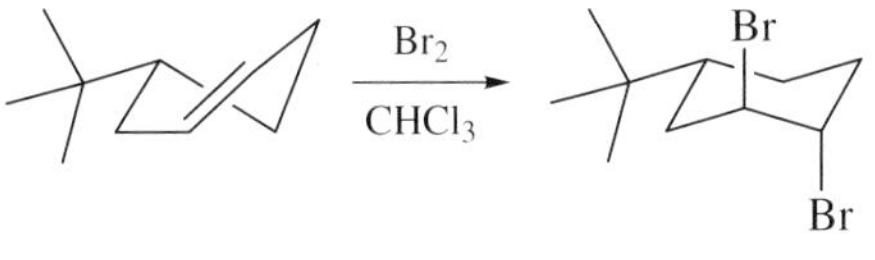

Br_2, $CHCl_3$ → Br, Br

Scheme 7.28

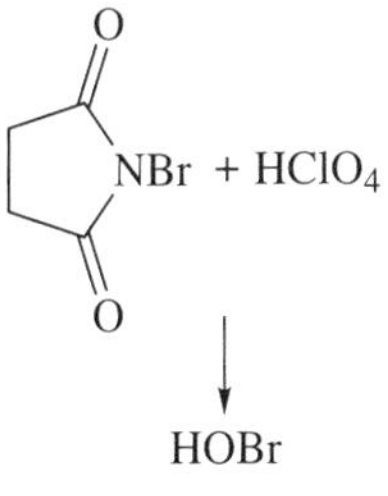

Hypohalite addition with hypochlorous acid (HOCl) or *t*-butyl hypochlorite (Me_3COCl) and methanol, or hypobromous acid (HOBr) (derived from *N*-bromosuccinimide and perchloric acid), gives the products of addition arising from attack by the halonium ion in the first step (Scheme 7.29).

NBr + $HClO_4$ → HOBr

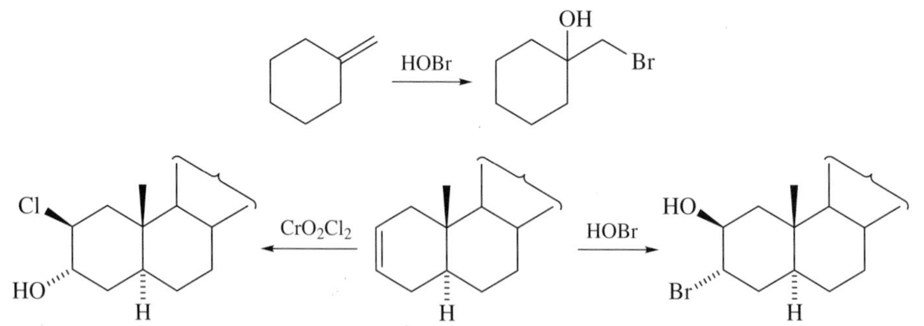

HOBr

CrO_2Cl_2

HOBr

Scheme 7.29

The regiochemistry of the addition of chromyl chloride contrasts with the addition of hypohalite. Thus propene gives 2-chloropropan-1-ol. The steroid example illustrates this difference (Scheme 7.29).

The regiospecific **halogenation of aromatic rings**, particularly bromination and iodination, has recently become more important because a number of methods of making aryl–carbon bonds are based on coupling reactions (*e.g.* the **Heck coupling**) The electrophilic halogenation of an aromatic ring can follow two distinct pathways. The first set of reactions involves the formation of a Wheland intermediate by electrophilic addition to the aromatic ring followed by elimination of a proton (see *Functional Group Chemistry*). The second pathway, typical of aromatic amides, first involves halogenation on the nitrogen atom, followed by an Orton rearrangement of the halogen to the *ortho* or *para* positions. Typical conditions for the bromination of a series of variously substituted aromatic compounds are given in the Table 7.3. Note the difference between the conditions that are required for activated aromatic rings (*e.g.* phenol) and those for deactivated rings (*e.g.* nitrobenzene).

Table 7.3 Bromination of aromatic compounds

Substrate	*Product*	*Conditions*
Benzene	Bromobenzene	Br_2 (neat), 65 °C, 15 min
Acetanilide	4-Bromoacetanilide	Br_2/glacial AcOH, room temp., 15 min
Aniline	2,4,6-Tribromoaniline	Br_2/glacial AcOH, room temp., 15 min
Phenol	4-Bromophenol	Br_2 (calc. amount), 0 °C
Benzoic acid	3-Bromobenzoic acid	Br_2/H_2O, sealed tube, 130 °C, 12 h
Nitrobenzene	1-Bromo-3-nitrobenzene	Br_2, Fe wire, 135–145 °C, 4 h

The monohalogenation of reactive aromatic rings can involve the introduction of blocking (protecting) groups or the use of very mild conditions. The preparation of 2-bromophenol is an example of this (Scheme 7.30).

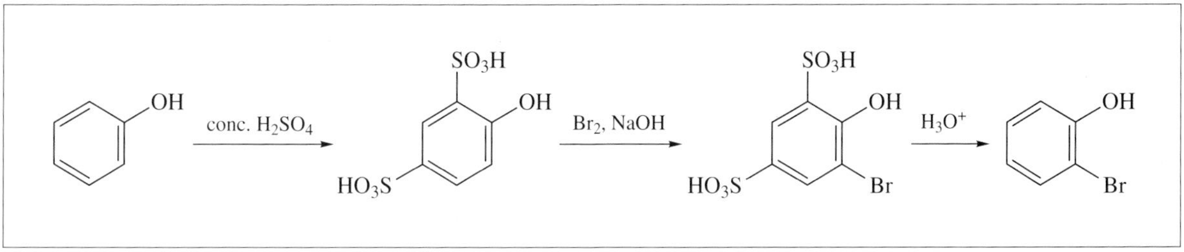

Scheme 7.30

Oxidative aromatic halogenation uses the bromide or iodide ion as the halogen source and an oxidant such as hydrogen peroxide or peracetic acid together with a catalyst in order to generate the electrophile (Scheme 7.31). For most iodinations the reactive species is not the iodonium ion. Iodinations usually require an oxidant and a mineral acid. The reactive species may be an iodosulfuric or iodonitric acid.

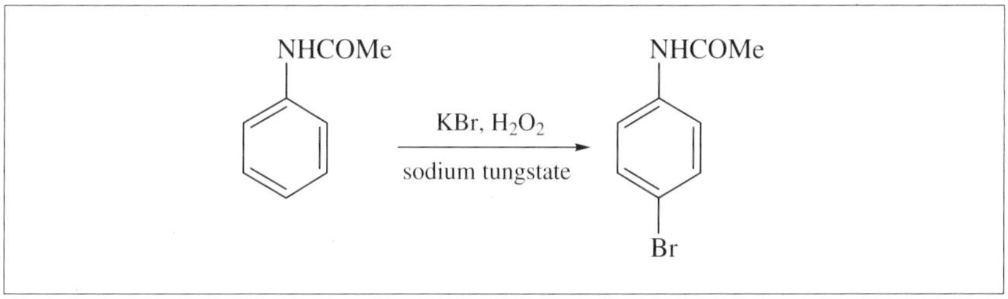

Scheme 7.31

Halogen can be introduced into an aromatic ring in place of various substituents. The diazotization of an aromatic amine gives an arenediazonium compound. The displacement of these with a halide in the presence of a copper salt (the **Sandmeyer reaction**) makes a useful method of introducing halogen. Heating the diazonium tetrafluoroborate provides a method for introducing fluorine (Scheme 7.32).

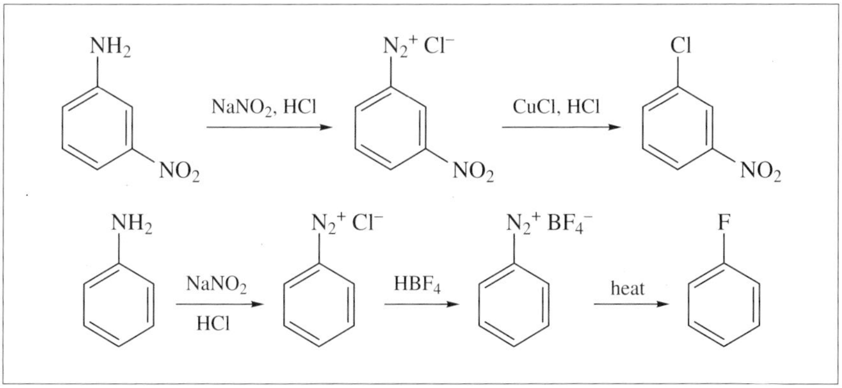

Scheme 7.32 The Sandmeyer reaction

Halogen exchange is usually difficult on aromatic rings. However, a halogen may be activated by an *ortho* or *para* nitro group. The low nucleophilicity of the fluoride in this context may be overcome by using potassium fluoride in DMF or by using silver fluoride.

7.3.3 Radical Methods

An alkene may be brominated in the allylic position by reaction with **N-bromosuccinimide** (NBS). The initial stage involves the formation of a resonance-stabilized allylic radical which abstracts a bromine atom from the NBS. Since the intermediate radical is delocalized over the adjacent double bond, there is the possibility of the formation of isomers in this reaction (Scheme 7.33).

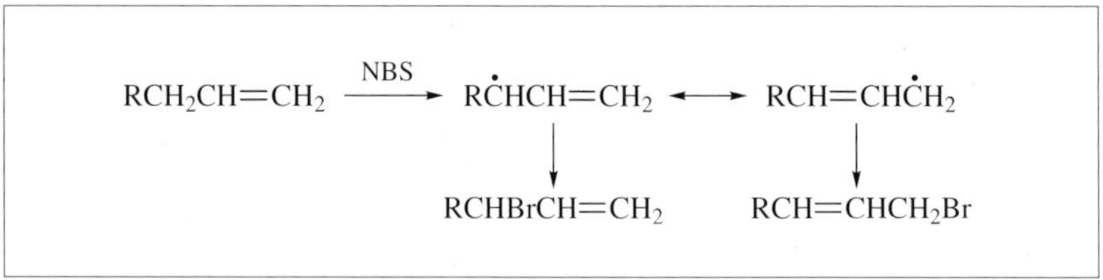

$$RCH_2CH{=}CH_2 \xrightarrow{NBS} R\dot{C}HCH{=}CH_2 \longleftrightarrow RCH{=}CH\dot{C}H_2$$

$$RCHBrCH{=}CH_2 \qquad RCH{=}CHCH_2Br$$

Scheme 7.33 The Hunsdiecker reaction

The bromodecarboxylation of the silver salt of a carboxylic acid (the **Hunsdiecker reaction**) provides a route for selectively introducing a halogen atom into a molecule (Scheme 7.34).

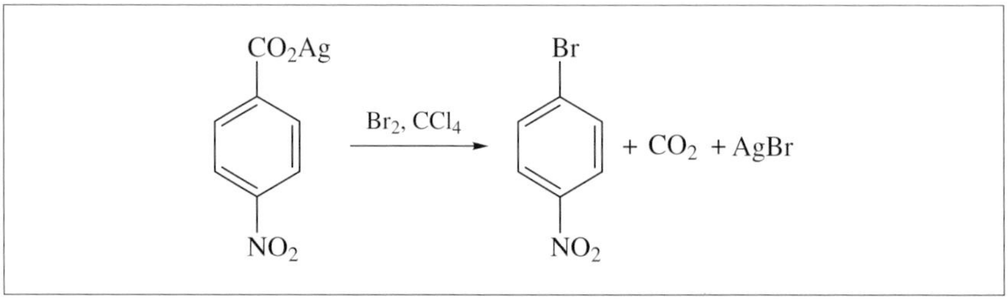

$$4\text{-}O_2NC_6H_4CO_2Ag \xrightarrow{Br_2,\ CCl_4} 4\text{-}O_2NC_6H_4Br + CO_2 + AgBr$$

Scheme 7.34

The consequences of radical and ionic conditions for bromination are illustrated by the bromination of 3-methoxycinnamic acid (Scheme 7.35).

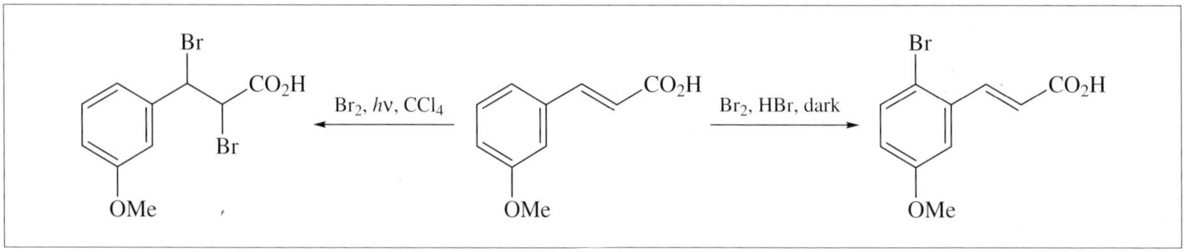

$$3\text{-}MeOC_6H_4CHBrCHBrCO_2H \xleftarrow{Br_2,\ h\nu,\ CCl_4} 3\text{-}MeOC_6H_4CH{=}CHCO_2H \xrightarrow{Br_2,\ HBr,\ dark} 2\text{-}Br\text{-}5\text{-}MeOC_6H_3CH{=}CHCO_2H$$

Scheme 7.35

Worked Problem 7.6

Q Propose a method for converting 5,6-dihydro-2*H*-pyran-2-one (**12**) into 2*H*-pyran-2-one (**13**).

12 → **13**

A The transformation involves the introduction of a double bond adjacent to an existing double bond. This may be achieved by the elimination of hydrogen bromide from an alkyl bromide, which is in turn prepared by the allylic bromination of the starting material by *N*-bromosuccinimide. The dehydrobromination uses a base such as triethylamine which is sufficiently bulky not to hydrolyse the lactone. The starting material **12** for the reaction can be prepared by the Prins reaction of but-3-enoic acid with formaldehyde:

CO_2H — HCHO, H_2SO_4 →

NBr → Br — Et_3N →

7.3.4 Fluorination

Organofluorine compounds have a number of interesting properties which distinguish them from other organohalogen compounds. Although the carbon–fluorine bond has a similar steric requirement to the carbon–hydrogen bond, it is considerably stronger and the fluorine atom is highly electronegative. The direct replacement of an alcohol with fluorine using hydrogen fluoride is of little use because of the problems associated with handling hydrogen fluoride. The hydrogen fluoride–pyridine complex is less corrosive and has been used for opening epoxides to make fluorohydrins (α-fluoro alcohols). The fluoride ion is solvated in aqueous or hydroxylic solvents and is then a relatively weak nucleophile. However, in dry aprotic solvents the fluoride ion is exposed and is a more powerful nucleophile. Thus tetrabutylammonium fluoride

($Bu_4N^+F^-$) can be used as a fluoride ion source in a solvent such as tetrahydrofuran, to displace a methanesulfonate or trifluoromethanesulfonate with inversion of configuration.

A useful reagent for the conversion of alcohols to fluoro compounds is **diethylaminosulfur trifluoride** (Et_2NSF_3, DAST). The reagent may function by forming a sulfimine that readily reacts with an alcohol. A second loss of fluoride converts the sulfur into a good leaving group, which is displaced by fluoride (Scheme 7.36). DAST can also be used to convert aldehydes and ketones to geminal difluorides and carboxylic acids to trifluoromethyl groups (Scheme 7.37).

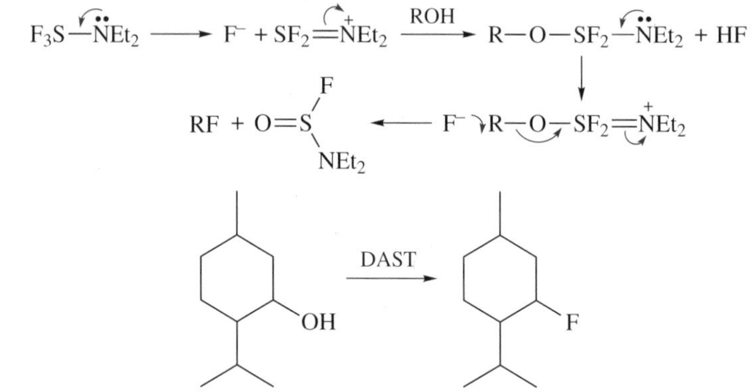

$$F_3S\text{—}\ddot{N}Et_2 \longrightarrow F^- + SF_2\text{=}\overset{+}{N}Et_2 \xrightarrow{ROH} R\text{—}O\text{—}SF_2\text{—}\ddot{N}Et_2 + HF$$

$$RF + O\text{=}S(F)(NEt_2) \longleftarrow F^- \; R\text{—}O\text{—}SF_2\text{=}\overset{+}{N}Et_2$$

DAST

OH F

Scheme 7.36

CO_2H DAST CF_3

Scheme 7.37

The trifluoromethyl group can also be introduced into a compound by an organometallic coupling procedure using copper or cadmium derivatives. These are prepared by the decomposition of sodium trifluoroacetate. Another way of introducing a trifluoromethyl group is as the trifluoromethyl radical ($CF_3\cdot$). This may be obtained by the decomposition of trifluoromethyl iodide or by electrolysis of trifluoroacetic acid. The radical can be trapped by an alkene (Scheme 7.38).

Because of the high electronegativity of fluorine, electrophilic fluorination is difficult to achieve. Selective fluorination to a carbonyl group can be achieved by reacting an enol ether with **xenon difluoride**. Other useful electrophilic fluorinating agents incorporate a powerful leaving group next to the fluorine. Examples are **trifluoromethyl hypofluorite** (CF_3OF) and a number of useful milder fluorinating agents containing

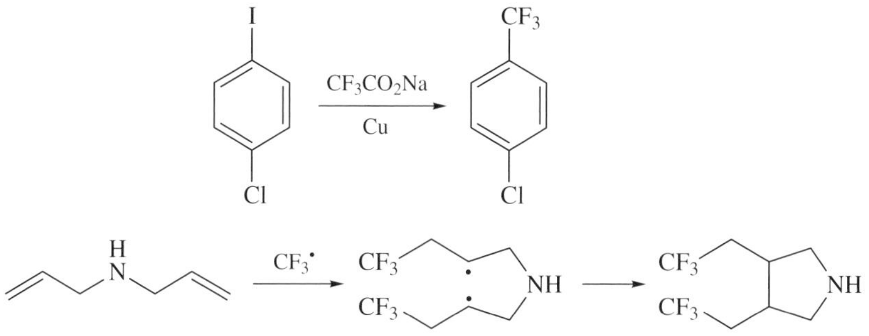

Scheme 7.38

the *N*-fluoro group such as *N*-fluorobenzene-1,2-disulfonimide (2-fluoro-1,3-benzodithiole 1,1,3,3-tetraoxide) and *N*-fluoropyridinium trifluoromethanesulfonate (Scheme 7.39).

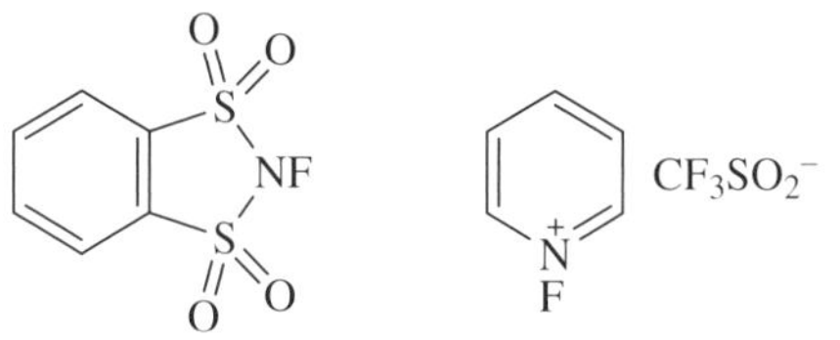

Scheme 7.39

Summary of Key Points

1. Oxidation reactions can be divided into dehydrogenation reactions, one and two-electron transfer reactions and oxygen donation reactions.

2. The dehydrogenation of ketones to α,β-unsaturated ketones extends the region of electron deficiency to the β-position.

3. The rate-determining step in the chromium(VI) oxide oxidation of alcohols is the fragmentation of the chromate(VI) ester.

4. The fragmentation of various derivatives of alcohols provides mild methods of oxidation.

5. The epoxidation and osmylation of alkenes can be adapted using chiral ligands to give chiral epoxides and diols.

6. Reduction reactions can be divided into hydrogenolysis, hydrogenation and oxygen abstraction reactions.

7. Hydrogenation reactions can be subdivided into catalytic methods, dissolving metal methods, hydride reactions and hydrogen transfer reactions.

8. The different groups of reagents show differing selectivities.

9. Hydroboration provides a useful method for the anti-Markownikoff hydration of an alkene.

10. Halogen can be introduced into a molecule by nucleophilic, electrophilic and radical methods.

Problems

7.1. Outline reactions by which cholest-5-ene (**14**) could be converted into compounds (a)–(g).

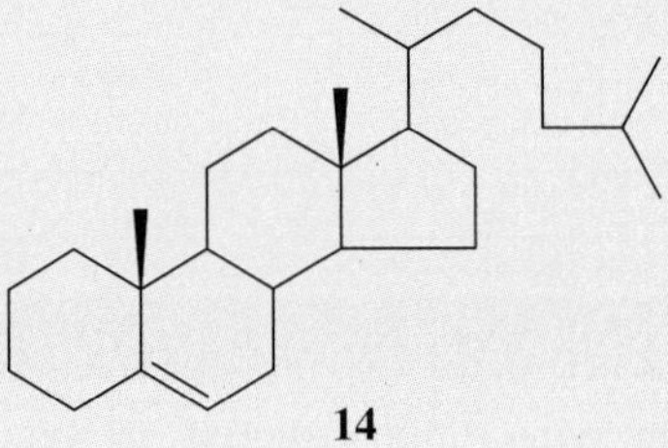

14

(a) OH OH

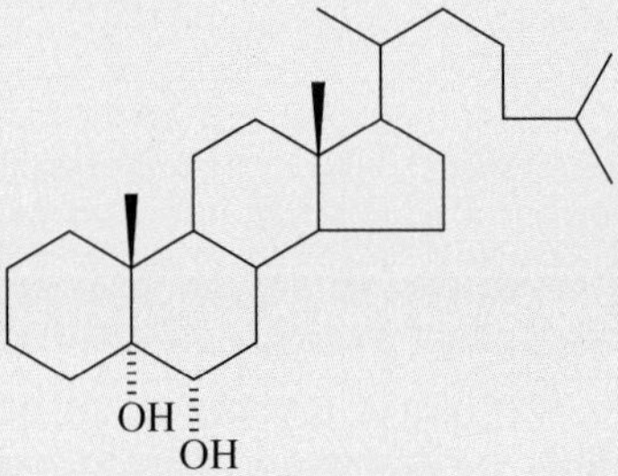

(b) O

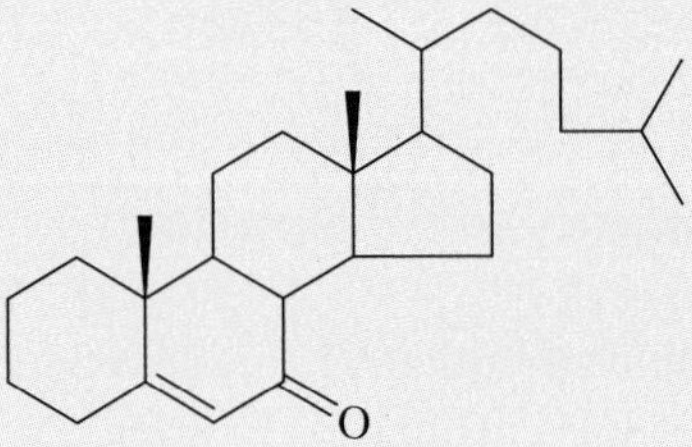

(c) OH OH

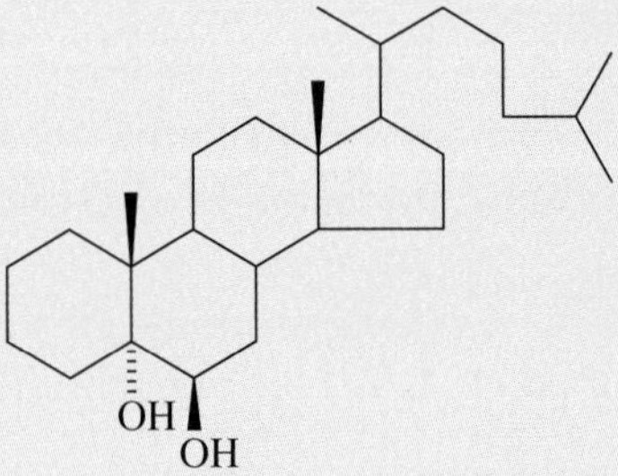

(d) OH

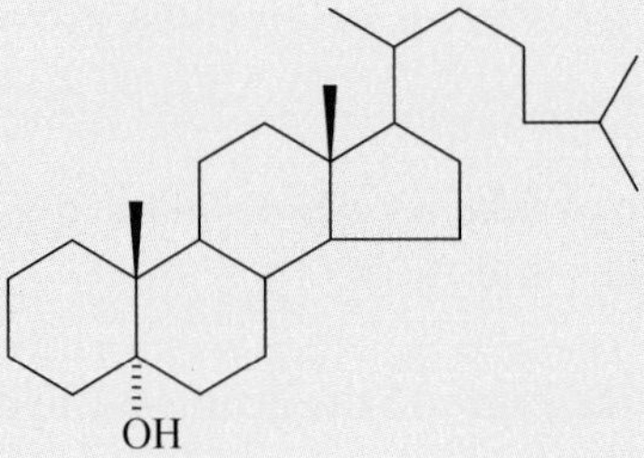

(e) H OH

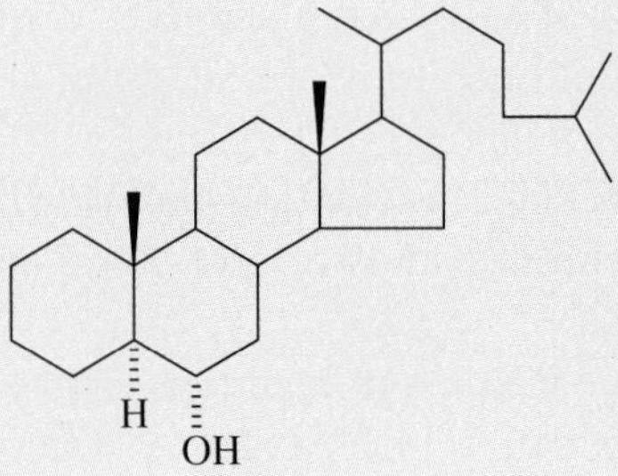

(f) 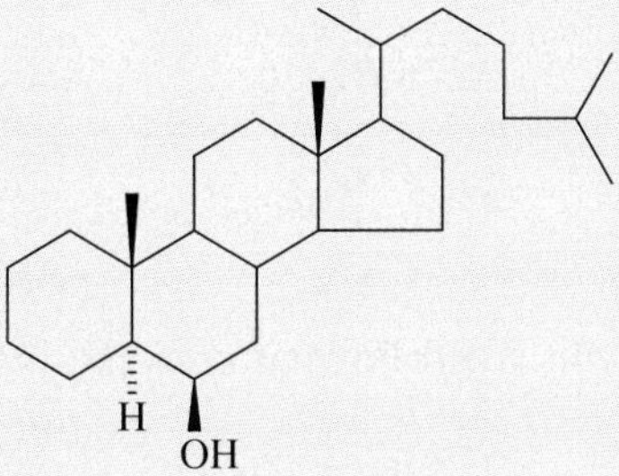 (g)

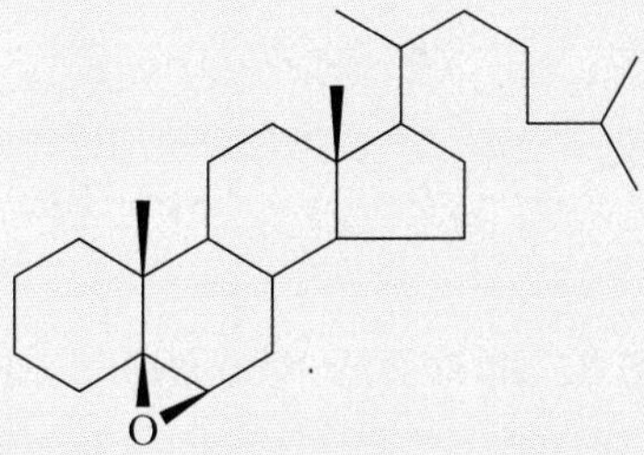

7.2. Show how would you convert the allylic alcohol **15** into compounds (a) and (b).

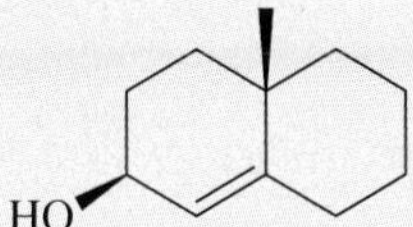

15

(a) 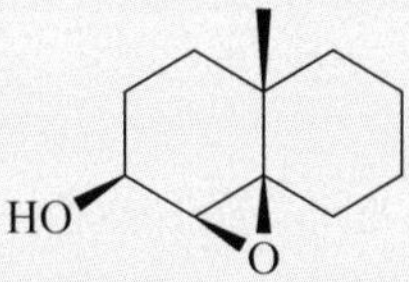 (b)

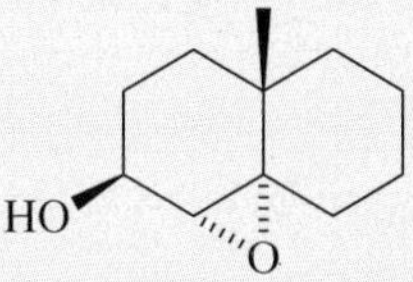

7.3. Show how you would convert $MeC{\equiv}CCH_2OH$ into (a) and (b).

(a) 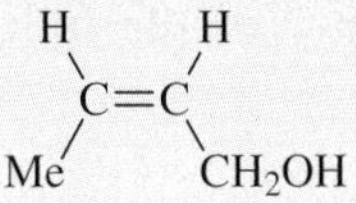 H, H / C=C / Me, CH_2OH (b) 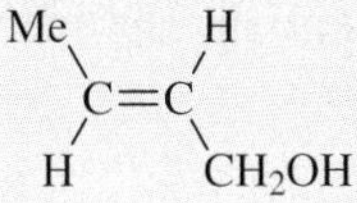 Me, H / C=C / H, CH_2OH

7.4. Show how you would convert $EtO_2CCH_2CH_2{}^{14}CO_2H$ into (a) and (b).

(a) $EtO_2CCH_2CH_2{}^{14}CH_2OH$ (b) $HOCH_2CH_2CH_2{}^{14}CO_2Et$

7.5. Show how would you convert methylcyclohexene (**16**) into compounds (a)–(d).

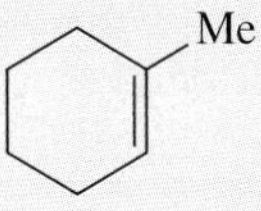

16 (a) 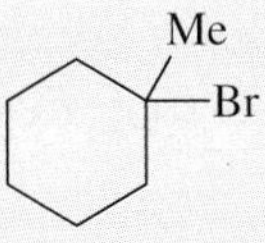 (b)

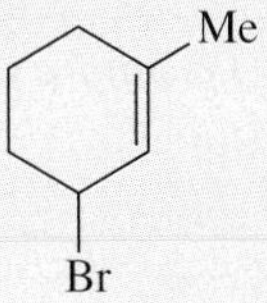

(c) 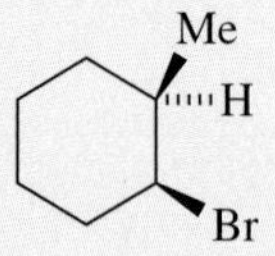 (d)

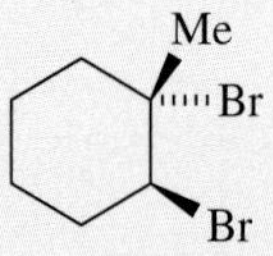

Me, H, Br — Me, Br, Br

7.6. Show how you would carry out the following conversions:

(a) 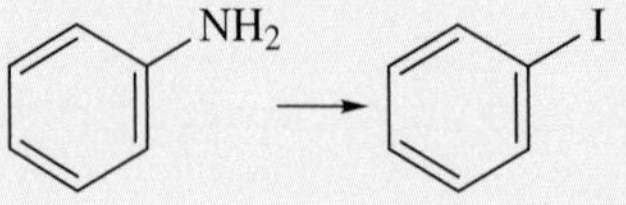 NH_2 → I (b) 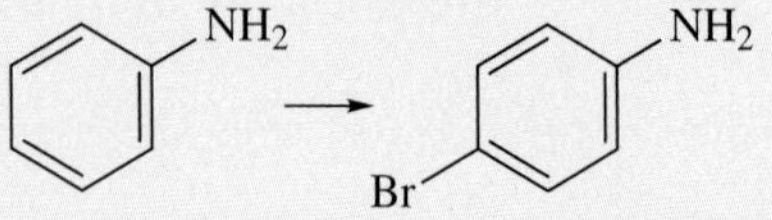 NH_2 → NH_2, Br

(c) 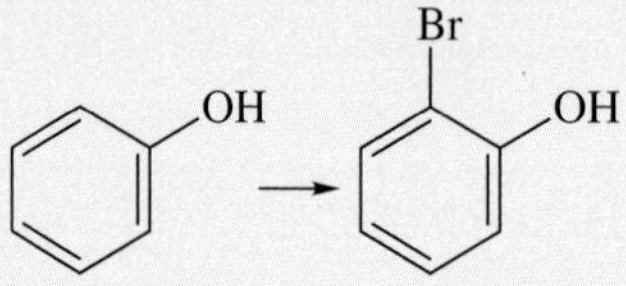 OH → Br, OH (d) 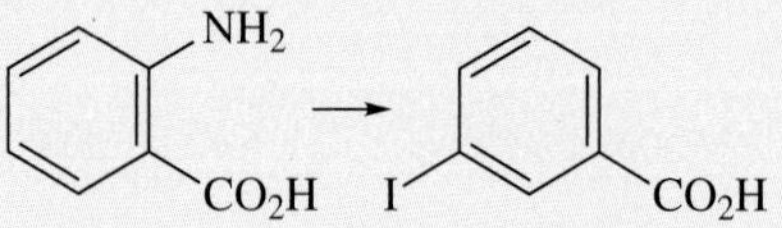 NH_2, CO_2H → I, CO_2H

7.7. Show how the following conversion might be carried out:

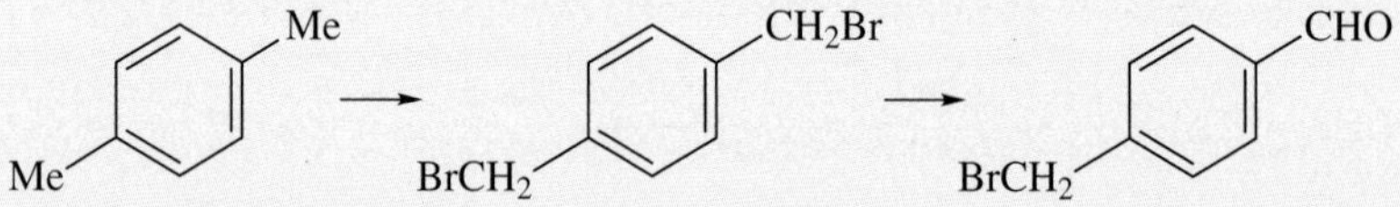

Me, Me → CH_2Br, $BrCH_2$ → CHO, $BrCH_2$

7.8. Comment on the stereochemistry of the following interconversions:

$CH(OH)CO_2H$ / CH_2CO_2H (−)-malic acid $\xrightarrow{PCl_5}$ $CH(Cl)CO_2H$ / CH_2CO_2H $\xrightarrow{Ag_2O}$ $CH(OH)CO_2H$ / CH_2CO_2H (+)-malic acid

Further Reading

A. J. Fatiadi, *The classical permanganate ion, still a novel oxidant in organic chemistry*, in *Synthesis*, 1987, 85.

F. A. Luzzio, *The oxidation of alcohols by modified oxochromium(VI)–amine reagents*, in *Org. React.*, 1998, **53**, 1.

A. J. Mancuso and D. Swern, *The oxidation of alcohols by activated dimethyl sulfoxide*, in *Synthesis*, 1981, 165.

G. R. Krow, *The Baeyer–Villiger oxidation of ketones and aldehydes*, in *Org. React.*, 1993, **43**, 251.

A. Pfenninger, *The asymmetric epoxidation of allylic alcohols, the Sharpless reaction*, in *Synthesis*, 1986, 89.

T. Katsuki and V. S. Martin, *Asymmetric epoxidation of allylic alcohols, the Katsuki–Sharpless reaction*, in *Org. React.*, 1996, **48**, 1.

H. C. Brown and S. Krishnamurthy, *Forty years of hydride reductions*, in *Tetrahedron*, 1979, **35**, 567.

E. R. H. Walker, *Functional group selectivity of complex hydride reducing agents*, in *Chem. Soc. Rev.*, 1976, **5**, 23.

G. W. Gribble, *Sodium borohydride in carboxylic acid media; a phenomenal reduction system*, in *Chem. Soc. Rev.*, 1998, **27**, 395.

S. Itsuno, *The enantioselective reduction of ketones*, in *Org. React.*, 1998, **52**, 395.

A. R. Pinder, *The hydrogenolysis of organic halides*, in *Synthesis*, 1980, 425.

I. Beletskaya and A. Pelter, *Hydroborations catalysed by transition metal complexes*, in *Tetrahedron*, 1997, **53**, 4957.

J. M. Hook and L. N. Mander, *Recent developments in the Birch reduction of aromatic compounds*, in *Nat. Prod. Rep.*, 1986, **3**, 35.

P. W. Rabideau and Z. Marcinow, *The Birch reduction of aromatic compounds*, in *Org. React.*, 1992, **42**, 1.

O. Mitsunobu, *The use of diethyl azodicarboxylate and triphenylphosphine in synthesis*, in *Synthesis*, 1981, 1.

D. L. Hughes, *The Mitsunobu reaction*, in *Org. React.*, 1992, **42**, 335.

J. Mann, *Modern methods for the introduction of fluorine into organic molecules*, in *Chem. Soc. Rev.*, 1987, **16**, 381.

J. A. Wilkinson, *Recent advances in the selective formation of the C–F bond*, in *Chem. Rev.*, 1992, **92**, 505.

J. R. Hanson, *Functional Group Chemistry*, Royal Society of Chemistry, Cambridge, 2001.

8
Protecting Groups

Aims

The aim of this chapter is to show how protecting groups may be used to mask the reactivity of a specific functional group. By the end of this chapter you should understand:

- The methods of introducing and removing ethers and esters for protecting alcohols, esters for protecting acids, amides and carbamates for protecting amino groups, and acetals for protecting carbonyl groups
- The role of protecting and activating groups in peptide synthesis
- The role of solid-phase methods in peptide synthesis and combinatorial chemistry

8.1 The Protection of Functional Groups

In the synthesis of a polyfunctional molecule there is often the need to introduce regiocontrol groups to favour reaction at a particular centre while other centres are left undisturbed. Protecting groups may be introduced to mask the reactivity of one centre while a reaction is carried out at another.

There are a number of important criteria which a protecting group must fulfil. First, it should be easily and quantitatively introduced under mild conditions which are specific for the sites in the molecule that need to be protected. Second, it must be stable to the reaction during which protection is required. Finally, it must be easily and quantitatively removed without causing disruption to the remainder of the molecule at the end of the reaction sequence. Furthermore, any residues from the protecting group should be easily separated. These criteria are often difficult to fulfil. Protecting groups have been classified into orthogonal (contrasting) sets based on their stability towards different groups of reagents.

The structures of many protecting groups are quite complex, and hence they are often abbreviated both in the text and in formulae. Some common abbreviations for protecting groups are given in Table 8.1.

Table 8.1 Some common abbreviations for protecting groups

Abbreviation	*Group*	*Abbreviation*	*Group*
Ac	Acetyl	MTM	(Methylthio)methyl
Bn	Benzyl	Ph	Phenyl
Boc	*t*-Butoxycarbonyl	SEM	2-(Trimethylsilyl)ethoxymethyl
Bu	Butyl	TBDMS	*t*-Butyldimethylsilyl
Bz	Benzoyl	TBTPS	*t*-Butyldiphenylsilyl
Cbz	Benzyloxycarbonyl (also Z)	TES	Triethylsilyl
Et	Ethyl	THP	Tetrahydropyranyl
Fmoc	9-Fluorenylmethoxycarbonyl	TIPS	Triisopropylsilyl
HOBT	1-Hydroxybenzotriazole	TMS	Trimethylsilyl
Me	Methyl	Tr	Triphenylmethyl (trityl)
MEM	2-Methoxyethoxymethyl	Troc	2,2,2-Trichloroethoxycarbonyl
MOM	Methoxymethyl	Ts	Toluene-*p*-sulfonyl
Ms	Methanesulfonate (mesylate)	Z	Benzyloxycarbonyl (also Cbz)

8.1.1 The Protection of Hydroxyl Groups

Several factors that contribute to the reactivity of a hydroxyl group need to be considered in selecting a protecting group. The hydrogen atom of the hydroxyl group is weakly acidic, and therefore reacts with strong bases and organometallic reagents such as Grignard reagents. Second, alcohols react with a number of reagents to form reactive esters that are intermediates in substitution and oxidation reactions. Third, the lone pair on the oxygen atom can react with mineral or Lewis acids. Furthermore, in many polyhydroxylic molecules it is necessary to differentiate between hydroxyl groups.

Simple methyl ethers, although blocking the hydroxylic hydrogen atom, are difficult to cleave. Therefore the structures of ethers have been modified to enhance their rate of cleavage. The stabilization of an incipient carbocation by an adjacent alkene or arene means that **allyl, benzyl** or **triphenylmethyl ethers** may be cleaved by acids. The increased steric bulk of the triphenylmethyl ether gives stereoselectivity to this protecting group, so that it will differentiate between primary and secondary alcohols. The conversion of the methyl ether to an acetal, as in a **methoxymethyl** or **tetrahydropyranyl ether**, increases the ease with which these groups may be removed under acidic conditions (Scheme 8.1).

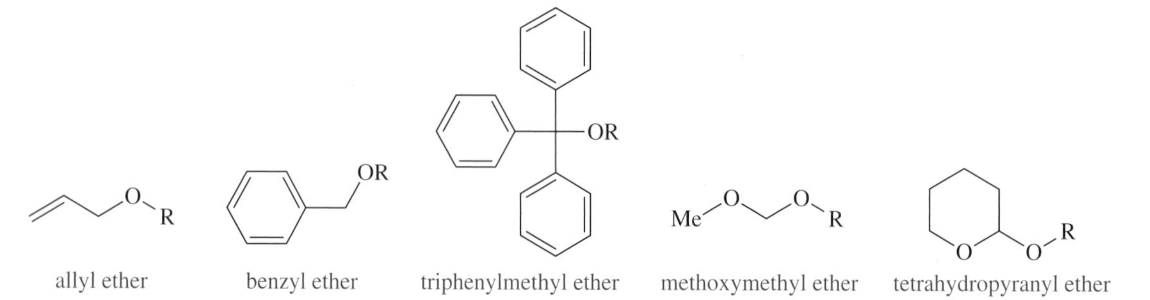

allyl ether | benzyl ether | triphenylmethyl ether | methoxymethyl ether | tetrahydropyranyl ether

Scheme 8.1 Ether protecting groups

The rate of cleavage may be increased by neighbouring group participation. For example, the rate of cleavage by a Lewis acid catalyst such as zinc chloride may be enhanced by incorporating a further oxygen atom into the protecting group, as in the **β-methoxyethoxymethyl** group (Scheme 8.2).

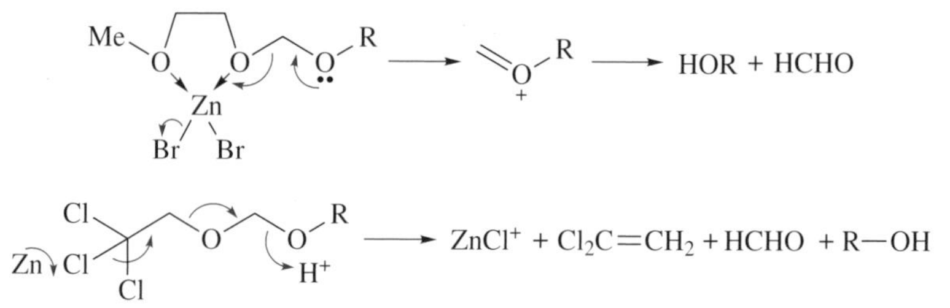

$\longrightarrow$ HOR + HCHO

$\longrightarrow$ $ZnCl^+$ + $Cl_2C{=}CH_2$ + HCHO + R—OH

Scheme 8.2

Fragmentation reactions can also provide selective methods for removing protecting groups. An example is the **β,β,β-trichloroethoxymethyl ether** (Scheme 8.2).

Protecting groups of this type are stable under conditions involving strong bases, hydride reducing agents, organometallic and many oxidizing reagents.

Silyl ethers provide a graded range of protecting groups, from trimethylsilyl through triethylsilyl, *t*-butyldimethylsilyl and *t*-butyldiphenylsilyl groups, in which steric factors increase the stability and decrease the sensitivity to hydrolysis (Scheme 8.3). The high bond enthalpy of the silicon–fluorine bond means that these groups are easily removed by fluoride ion.

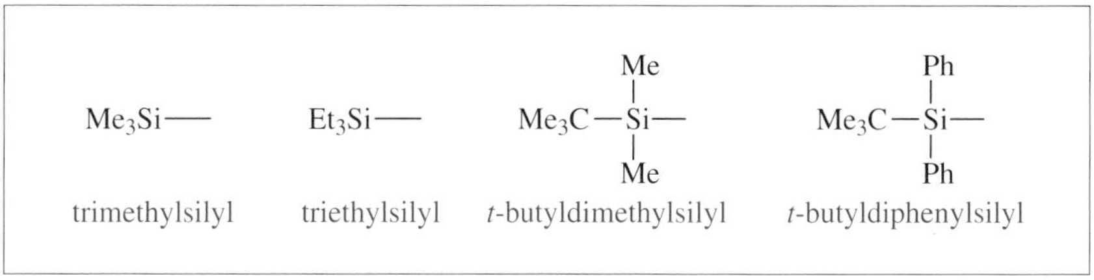

$Me_3Si—$ trimethylsilyl; $Et_3Si—$ triethylsilyl; $Me_3C—Si(Me)_2—$ *t*-butyldimethylsilyl; $Me_3C—Si(Ph)_2—$ *t*-butyldiphenylsilyl

Scheme 8.3 Silyl ether protecting groups

Esters provide good protecting groups for reactions under acidic conditions because the interaction of the electron-withdrawing carbonyl group of the ester with the lone pair on the ester oxygen renders this atom less basic. **Acetate** and **benzoate** esters are commonly used as protecting groups against oxidation reactions. Various modifications have been made to the ester grouping, such as the insertion of an additional oxygen atom to make a carbonate protecting group in order to enhance their removal (Scheme 8.4).

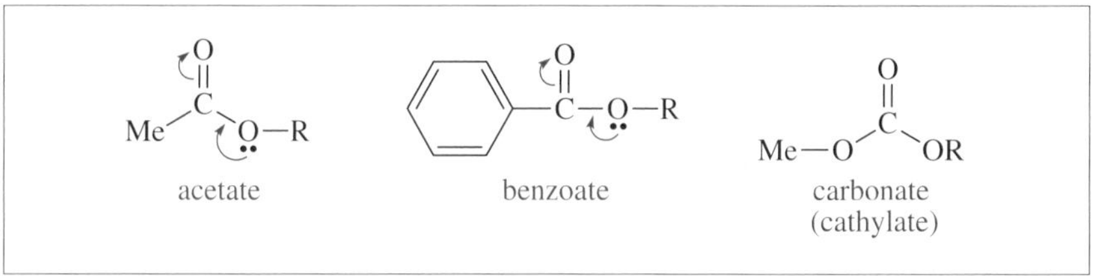

acetate; benzoate; carbonate (cathylate)

Scheme 8.4 Ester protecting groups

The selective protection of 1,2-diols in sugars, for example, has been developed using cyclic acetals based on acetone (propanone) (**acetonides**) and benzaldehyde (**benzylidene**) derivatives. There is an interesting contrast between these two protecting groups in the protection of the triol glycerol. Glycerol forms a 1,3-benzylidene derivative with benzaldehyde and a 1,2-acetonide with acetone (Scheme 8.5). 1,2-Diols may also be protected as their cyclic carbonates, which can be prepared with phosgene (carbonyl dichloride, $COCl_2$) or, better, the less toxic triphosgene [bis(trichloromethyl) carbonate, $CCl_3OC(O)OCCl_3$].

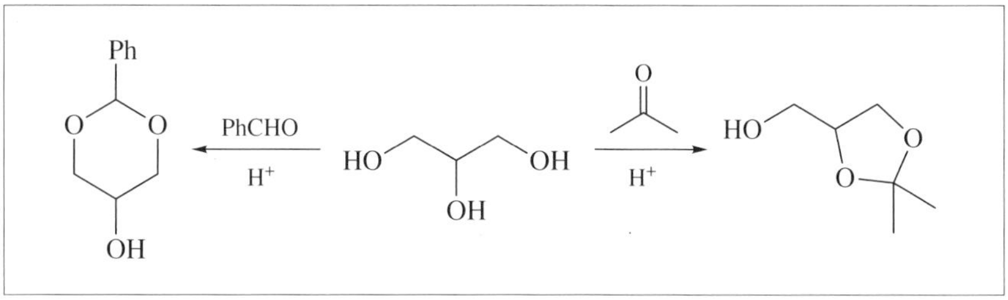

PhCHO, H^+; H^+

Scheme 8.5

Worked Problem 8.1

Q Propose suitable protecting groups to enable the C-3 hydroxyl group of 2-deoxyribose (**1**) to be oxidized by chromium(VI) oxide in pyridine.

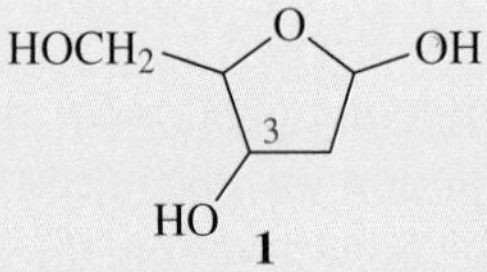

HOCH$_2$ O OH 3 HO **1**

A The C-1 hydroxyl group is part of an acetal and may be protected by conversion to the methyl ether by methanol and hydrochloric acid. The primary C-5 hydroxyl group may be protected by a bulky protecting group such as triphenylmethyl or *t*-butyldimethylsilyl. A bulky protecting group would leave the secondary C-3 alcohol exposed for the reaction. The protecting groups could then be removed by treatment with acid.

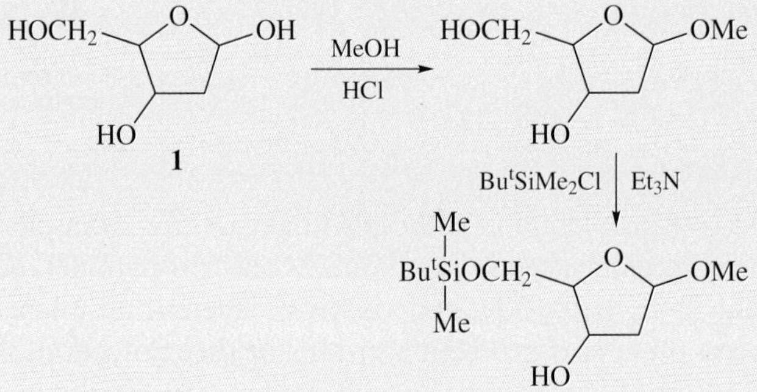

HOCH$_2$ O OH HO **1** → MeOH, HCl → HOCH$_2$ O OMe HO → Bu^{t}SiMe$_2$Cl, Et$_3$N → Bu^{t}Si(Me)$_2$OCH$_2$ O OMe HO

8.1.2 Carboxylic Acids

The acidic hydrogen of carboxylic acids needs to be protected in order to prevent it reacting with basic reagents. This protection may be carried out with a **methyl ester**. Such methyl esters may be prepared with methanol and the acid in the presence of a catalyst or with diazomethane. However, simple alkaline hydrolysis of esters can present a problem because of the vigorous conditions involved. The hydrolysis of a carboxylic acid ester by base involves addition to the carbonyl group and the formation of a tetrahedral intermediate. A number of alternative reagents have been developed, based on alkyl–oxygen fission. These include trimethylsilyl iodide, lithium iodide or lithium propanethiolate. In these reagents a hard acid (the silicon or lithium) coordinates with the oxygen and the softer base (iodide or a thiol) attacks the carbon of the alkyl group.

The *tert*-butyl ester is a useful protecting group for carboxylic acids because of the steric hindrance it provides against nucleophilic attack on the carbonyl group. This ester is sensitive to cleavage under mild acid-catalysed conditions to release the gaseous isobutene. Benzyl esters may also be removed by acid-catalysed or hydrogenolytic methods.

Some esters that can be removed by fragmentation reactions are useful protecting groups. These are exemplified by the **β,β,β-trichloroethyl ester,** which may be reductively cleaved with zinc in an elimination reaction, and the **β-(trimethylsilyl)ethoxymethyl ester**, which undergoes fragmentation with a fluoride ion (Scheme 8.6).

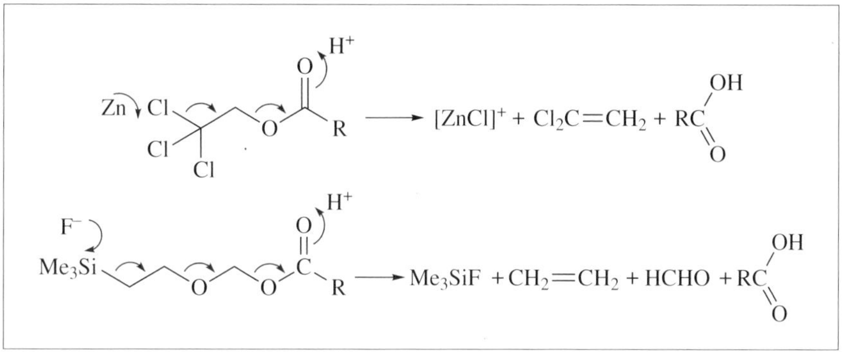

$$\text{Zn} + \text{Cl}_3\text{C–CH}_2\text{–O–C(=O)R} \xrightarrow{\text{H}^+} [\text{ZnCl}]^+ + \text{Cl}_2\text{C=CH}_2 + \text{RCO}_2\text{H}$$

$$\text{F}^- + \text{Me}_3\text{Si–CH}_2\text{CH}_2\text{–O–CH}_2\text{–O–C(=O)R} \xrightarrow{\text{H}^+} \text{Me}_3\text{SiF} + \text{CH}_2\text{=CH}_2 + \text{HCHO} + \text{RCO}_2\text{H}$$

Scheme 8.6

8.1.3 Protection of Carbonyl Groups

The electron-withdrawing properties of the carbonyl group make it sensitive to nucleophilic addition and render the adjacent C–H acidic. Hence it can be necessary to mask the reactivity of a carbonyl group during the course of a synthesis. This can be achieved by converting the carbonyl group to an acetal. Although dimethoxy acetals can be prepared, particularly from aldehydes, **1,3-dioxolanes**, derived from ethane-1,2-diol, provide the most useful protecting groups. Dioxolanes are stable to alkaline conditions and to many organometallic reagents, They are formed and cleaved under acid-catalysed conditions. Because a carbonyl group is converted to a tetrahedral centre, there is an increase in steric bulk associated with the formation of a 1,3-dioxolane. This can provide some selectivity in, for example, the protection of 4-oxoisophorone, where only the less-hindered carbonyl group forms a 1,3-dioxolane with ethane-1,2-diol.

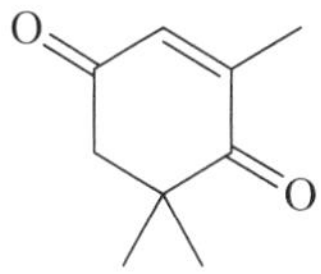

4-oxoisophorone

Extra functionality has been introduced into the dioxolane group to allow fragmentation reactions to be used in the deprotection step. For example, 1-bromo-2,3-dihydroxypropane forms a dioxolane as shown in Scheme 8.7. Regeneration of the ketone may be brought about by a reductive elimination with zinc.

$$R_2C{=}O + HOCH_2CH_2OH \longrightarrow R_2C(OCH_2CH_2O)$$

$$\text{Zn–BrCH}_2\text{CH}_2\text{(O}_2\text{CR}_2\text{)} + H^+ \longrightarrow [ZnBr]^+ + CH_2{=}CHCH_2OH + RC(O)R$$

Scheme 8.7

dithioacetal

Treatment of an aldehyde or ketone with a thiol in the presence of an acid catalyst gives a **dithioacetal**. Cyclic dithioacetals are formed from ethanedithiol or propane-1,3-dithiol. The equilibrium is normally in favour of the thioacetal. However, the deprotection reaction is more difficult, and usually requires salts of metals such as mercury or cadmium which form strong metal–sulfur bonds.

Worked Problem 8.2

Q Propose a suitable protecting group for the carbonyl group of 1-bromobutan-3-one which would allow the nucleophilic substitution of the bromine by an alkyne anion:

$$BuC{\equiv}C^- Na^+ + BrCH_2CH_2\overset{O}{\overset{\|}{C}}Me \rightleftharpoons BuC{\equiv}CCH_2CH_2\overset{O}{\overset{\|}{C}}Me$$

A A base-stable protecting group is required which would protect the carbonyl group against nucleophilic addition. The ethylene ketal would be a suitable protecting group:

$$BuC{\equiv}C^- Na^+ + BrCH_2CH_2C(Me)(OCH_2CH_2O) \longrightarrow BuC{\equiv}CCH_2CH_2C(Me)(OCH_2CH_2O)$$

$$\xrightarrow{H_3O^+} BuC{\equiv}CCH_2CH_2C(O)Me$$

8.1.4 Amino Groups

An amino group has both basic and nucleophilic properties. Amines form complexes with metal salts and are susceptible to oxidation and the N–H is sufficiently acidic to react with organometallic reagents. Amines are protected both by alkylation and acylation. The formation of an amide from an amine diminishes the nucleophilicity of the nitrogen, as in the acetamide in Scheme 8.8. However, if the amine involved is part of an amino acid that is incorporated into a peptide, special protecting groups have to be devised that may be removed in the presence of the amide peptide bond.

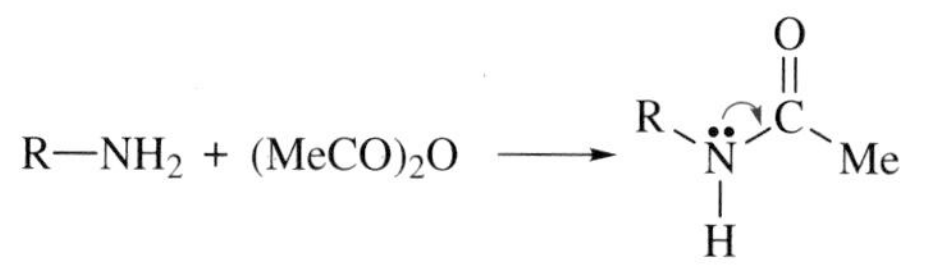

$$R{-}NH_2 + (MeCO)_2O \longrightarrow R{-}NH{-}C(=O){-}Me$$

Scheme 8.8

The introduction of alkyl groups on to the nitrogen of an amine blocks the N–H group. However, the groups that are introduced must be selectively removed. A benzyl group is a suitable protecting group in this context because it can be removed by hydrogenolysis. A primary amine can be protected by incorporating the nitrogen into a pyrrole ring (Scheme 8.9). The amine can be regenerated at the end of a reaction sequence by deprotection with hydroxylamine hydrochloride, which forms a dioxime with the diketonic component of the protecting group.

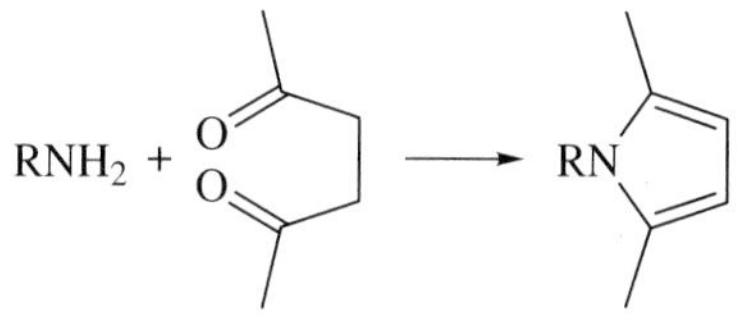

$$RNH_2 + \text{hexane-2,5-dione} \longrightarrow RN\text{(2,5-dimethylpyrrole)}$$

Scheme 8.9

The protection of amines as amides reduces the nucleophilic character of the nitrogen atom. Thus an aromatic amine, such as aniline, is converted to the amide, acetanilide, prior to aromatic substitution.

8.2 Peptide Synthesis

The protection of the amino group in an amino acid is necessary to ensure a regiospecific coupling between amino acids in peptide synthesis. At the same time, the carboxyl group of this amino acid is activated towards nucleophilic attack and the carboxyl group of the other amino acid is protected (Scheme 8.10).

The protecting groups must be removed under conditions that are sufficiently mild to avoid cleaving the peptide bond and disturbing the chiral centres of the amino acids. Typical protecting groups contain a carbamate unit, which combines the low nucleophilicity of an amide with an easy decarboxylation in the deprotection step.

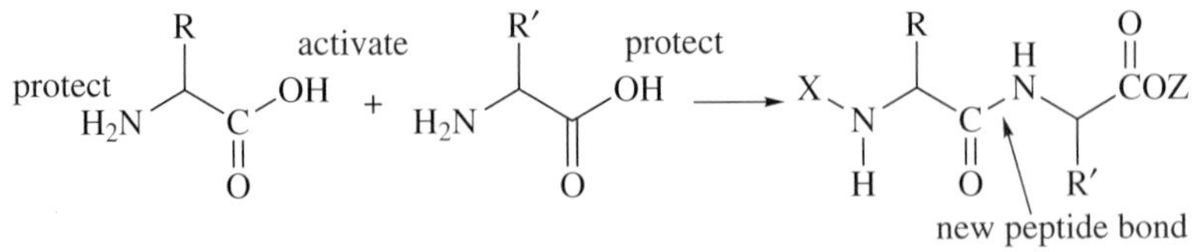

Scheme 8.10

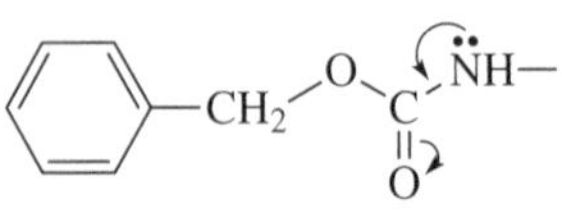

benzyloxycarbonyl (Z or Cbz)

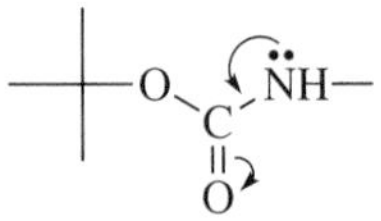

t-butoxycarbonyl (Boc)

The **benzyloxycarbonyl** (Z or Cbz) and ***t*-butoxycarbonyl** (Boc) protecting groups are examples of this type of protecting group. The benzyloxycarbonyl group is introduced using benzyl chloroformate, whilst the *t*-butoxycarbonyl group is introduced *via* di-*t*-butyl carbonate or an oxime derivative. These protecting groups are stable to basic conditions, but may be removed either with acid or, in the case of the benzyl group, by hydrogenolysis. In each case a carbamic acid [HOC(O)NHR] is formed, which readily loses carbon dioxide.

Whereas these protecting groups are removed under acidic conditions, a variant, the **9-fluorenylmethoxycarbonyl** (Fmoc) protecting group, is acid stable and base labile. It fragments in the presence of a base such as piperidine (Scheme 8.11).

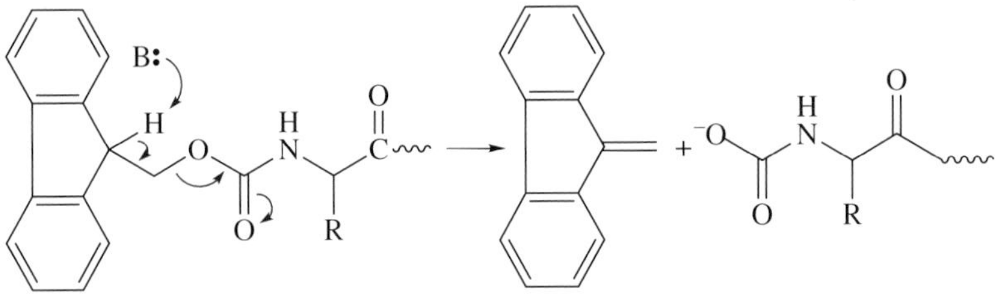

Scheme 8.11

The carboxyl group at which the peptide bond is to be formed is activated for nucleophilic attack. The sensitivity of the carbonyl group of a carboxylic acid to nucleophilic attack is reduced by the presence of the hydroxyl group, particularly if the nucleophile is sufficiently basic to form the carboxylate anion. Consequently, activating groups are introduced to enhance the sensitivity of the carbonyl group to addition reactions, to modify the effect of the lone pairs, and to remove the acidic hydrogen.

The conversion of an acid to the acyl chloride or acyl azide gives a good leaving group for amide formation *via* a tetrahedral intermediate. However, the conditions for making the acyl chloride are quite vigorous and lead to racemization of the amino acids, while the preparation of azides involves reacting the ester of the acid with hydrazine and then reacting the hydrazide with nitrous acid.

In an acid anhydride, the resonance involving one carbonyl deactivates the lone pair of the singly bonded oxygen, exposing the other carbonyl to nucleophilic attack. In a symmetrical anhydride this results in the loss of one carboxylate unit. Consequently, in the case of protected amino acids, various **mixed anhydrides** have been prepared. A mixed carbonic anhydride can be prepared from ethyl chloroformate (Scheme 8.12). The presence of two oxygen substituents directs the nucleophile to the carbonyl of the amino acid. A mixed anhydride with diphenylphosphinic acid plays a similar role (Scheme 8.12).

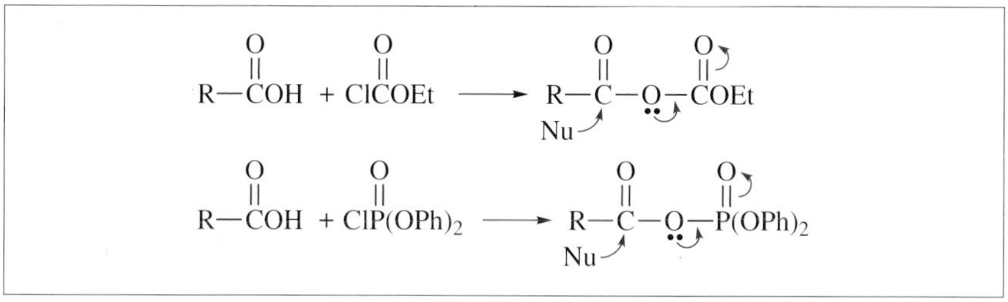

$$R{-}C(=O)OH + ClC(=O)OEt \longrightarrow R{-}C(=O){-}O{-}C(=O)OEt \quad (Nu)$$

$$R{-}C(=O)OH + ClP(=O)(OPh)_2 \longrightarrow R{-}C(=O){-}O{-}P(=O)(OPh)_2 \quad (Nu)$$

Scheme 8.12

Dicyclohexylcarbodiimide (DCC) is a well-established carboxyl-activating group. The central diimide carbon is very electron deficient, and hence readily adds the carboxyl oxygen to give an *O*-acylurea (Scheme 8.13). The resonance involving the remaining amide group deactivates the oxygen lone pairs sufficiently to allow nucleophilic addition to the carboxyl carbonyl group. Collapse of the tetrahedral intermediate leads to the formation of a urea from the DCC as well as to the peptide.

A number of **reactive esters** have been prepared from protected amino acids to help the formation of the peptide bond. Some examples (Scheme 8.14) are the 4-nitrophenyl ester, the pentafluorophenyl ester,

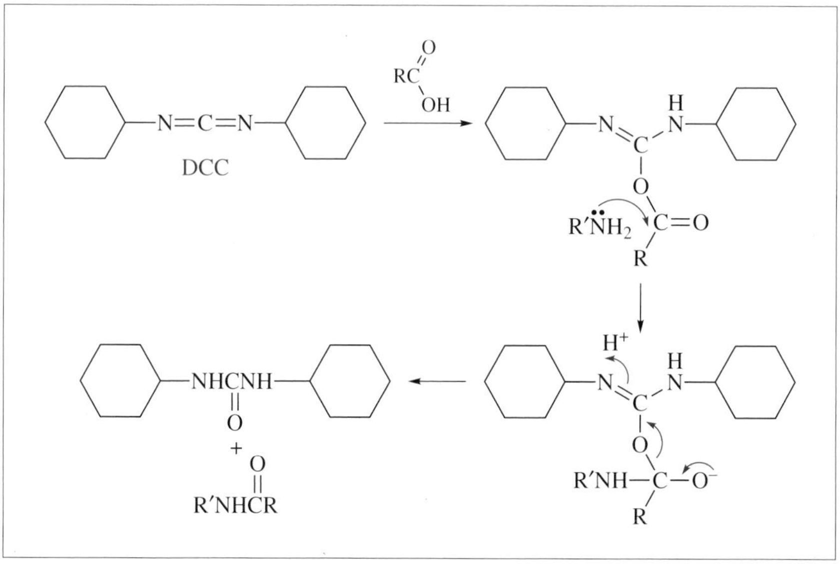

C₆H₁₁–N=C=N–C₆H₁₁ (DCC) + RC(=O)OH ⟶ C₆H₁₁–N=C(–NH–C₆H₁₁)–O–C(=O)R; R′NH₂

⟶ (H⁺) C₆H₁₁–N=C(–NH–C₆H₁₁)–O–C(O⁻)(R)–NHR′ ⟶ C₆H₁₁–NHC(=O)NH–C₆H₁₁ + R′NHC(=O)R

Scheme 8.13

the 2,4,5-trichlorophenyl ester, the *N*-hydroxysuccinimide ester and the 1-hydroxybenzotriazole ester. In each case the reactivity may be associated with a resonance interaction between the lone pair on the oxygen of the ester and the esterifying group.

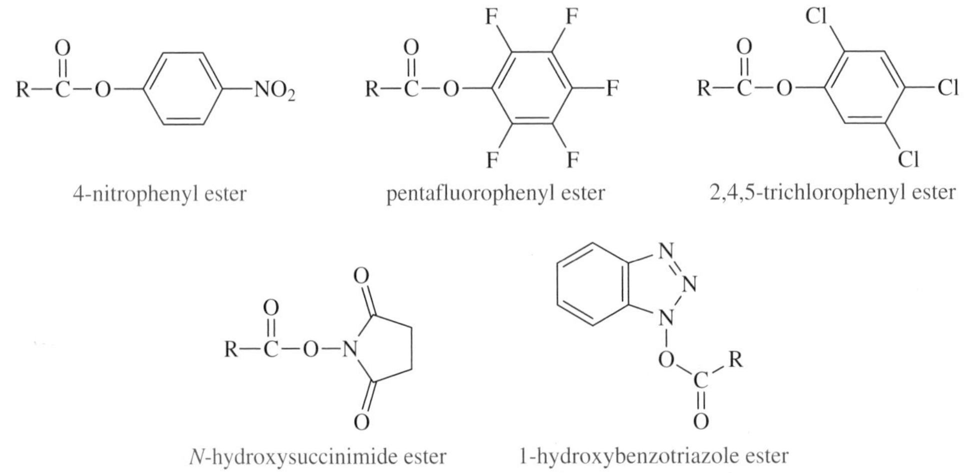

4-nitrophenyl ester pentafluorophenyl ester 2,4,5-trichlorophenyl ester

N-hydroxysuccinimide ester 1-hydroxybenzotriazole ester

Scheme 8.14 Activating esters

Protection of the remaining carboxyl of the amino acid, the Z group, usually involves an ester. This may be a bulky ester such as a **benzyl ester**, to sterically hinder the approach of a nucleophile to this centre.

The synthesis of a peptide involves a sequence of repetitive operations involving coupling and purification steps. In order to force reactions to completion, an excess of reactants may be used. This can provide a serious purification problem. A solid-phase peptide synthesis, known after its inventor as the **Merrifield method**, uses a resin as the Z group (Scheme 8.15). The resin has attached to it a number of benzyl groups to which the first amino acid of the chain is attached. The second protected amino acid is coupled to this. The excess reactant can be washed from the resin and the protecting groups removed, and their residue removed by washing the solid phase to which the developing peptide chain is attached. Further coupling and deprotection steps can be carried out. Each time the peptide remains attached to the resin, but the residues of the reactants and reagents are easily separated by washing. When the peptide is finally complete, it can be removed from the resin by treatment with hydrobromic acid (Scheme 8.15). This solid phase approach has revolutionized peptide synthesis and it has been automated.

8.3 Combinatorial Synthesis

The development of rapid enzymatic assays by the pharmaceutical indus-

add

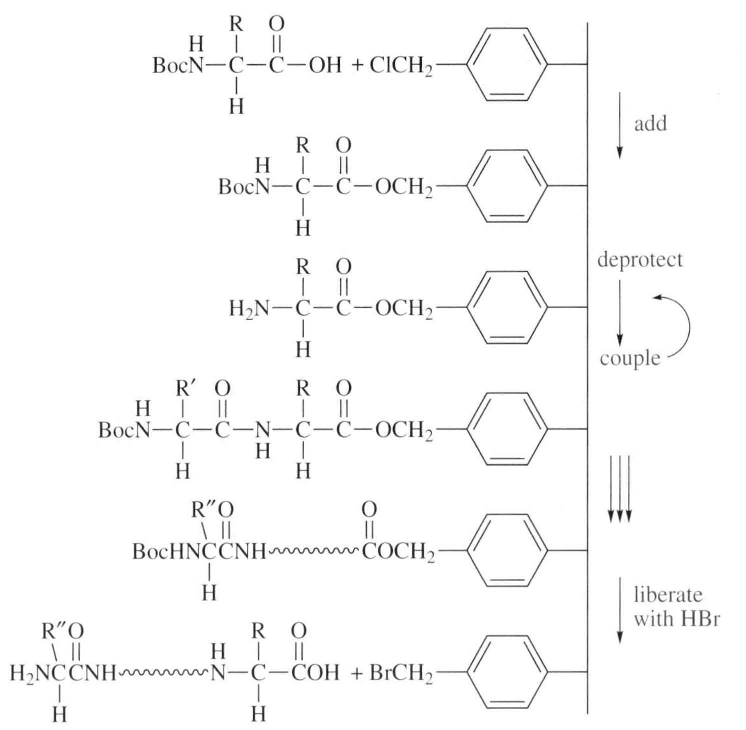

deprotect

couple

liberate with HBr

Scheme 8.15

try in place of the older whole-organism bioassays has meant that many more compounds can be screened very rapidly on a smaller scale. Combinatorial methods of synthesis were developed to fulfil the need of high-throughput screening. Many synthetic methods are quite general and have been optimized, and hence they may be applied to a range of compounds in parallel. At its simplest, combinatorial chemistry is doing just that. Consider making an amide from an amine and an acyl chloride. If an array of four acyl chlorides are reacted with four amines in a set of reaction vessels in parallel under standard conditions, 16 amides will be produced which can be subjected to bioassay. Since the conditions for each reaction and its work-up are essentially identical, the process can be miniaturized and automated. Indeed, it can be made to work in 8×12 (96) well plates. In a simple example, a small library of 20 substituted aminothiazoles was obtained by the combination of five different thioureas and four α-bromo ketones (Scheme 8.16).

There are restrictions to the application of solution phase chemistry because of the need to use, and subsequently separate, the excess reagent required to drive reactions to completion. The build-up of impurities in a multi-step synthesis then becomes a problem. However, these may be overcome by attaching the starting material through a linker, such as a phenoxybenzyl alcohol, to an insoluble resin or bead as in the Merrifield

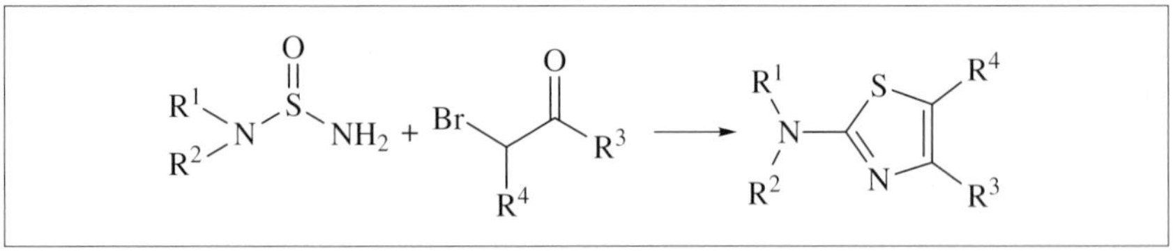

Scheme 8.16

method for peptide synthesis. The reactions are driven to completion by adding excess reagent, which can then be removed by filtering the bead containing the product and washing it thoroughly before the next step in the synthesis. A typical example of this type of synthesis involved the preparation of benzodiazepines from resin-bound amino acid esters and substituted anthranilic acids (Scheme 8.17). In this example the amino acid esters were attached to the solid support and acylated with the anthranilic acids. Base-catalysed hydrolysis and lactamization led to the formation of the benzodiazepine ring. This was then alkylated and the products were cleaved from the resin.

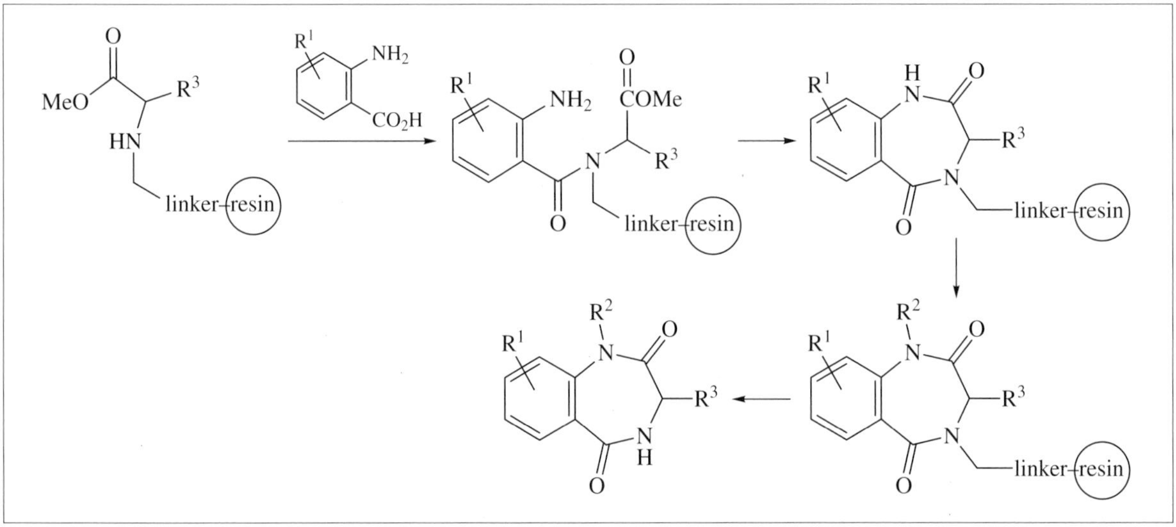

Scheme 8.17

Once the substrates are attached to the beads, multiple-step syntheses can be achieved by what is known as "split and mix" methodology. Suppose two different starting materials A and B are each attached to a different group of beads. The beads may then be mixed and split into two again, and reacted with two different reagents C and D to give four possible combinations of products:

C	C	D	D
A	B	A	B

The beads may again be thoroughly washed, recombined and split into a further group of two and reacted with reagents E and F to give a further range of products:

E	E	E	E	F	F	F	F
C	C	D	D	C	C	D	D
A	B	A	B	A	B	A	B

The compounds can then be removed from the beads. In this simple example, a library of eight compounds ($2 \times 2 \times 2$) is produced. However, libraries of 500 compounds ($10 \times 10 \times 5$) and even much larger can be generated quite easily. In the example shown in Scheme 8.17 a library of 2508 compounds was obtained from 19 amino acids, 12 anthranilic acids and 11 alkylating agents. Once the compounds have been released from the beads and biologically active compounds obtained, there can be a problem of identification. In this context it is important to remember that multiple copies of just one compound are attached to each bead, and not all needs to be used in the bioassay. Although large numbers of compounds can be generated this way, it is important to realise that the design of the library must reflect a careful consideration of the biological target.

Summary of Key Points

1. Methyl ethers are difficult to hydrolyse and hence modified ethers such as methoxyethoxymethyl, tetrahydropyranyl and triphenylmethyl are used to protect alcohols.

2. Silyl ethers are useful protecting groups for alcohols. Steric factors can be used to differentiate between various alcohols.

3. Esters can be used to protect alcohols. Substituents have been introduced to facilitate their cleavage.

4. Acetonide and benzylidene protecting groups are specific for diols.

5. Substituted alcohols can be used to protect carboxylic acids as their esters.

6. Acetals and thioketals can be used to protect carbonyl groups.

7. Amine protecting groups such as benzyloxycarbonyl (Cbz or Z) and *t*-butoxycarbonyl (Boc) are widely used in peptide synthesis.

8. Solid phase methods have significant advantages, particularly in purification, in both peptide synthesis and combinatorial chemistry.

Problems

8.1. Suggest suitable protecting groups for the following transformations:

$HOCH_2CH_2CH(OH)CH_2OH \rightarrow$ (a) $HO_2CCH_2CH(OH)CH_2OH$

$HOCH_2CH_2CH(OH)CH_2OH \rightarrow$ (b) $HOCH_2CH_2C(=O)CH_2OH$

8.2. Suggest suitable protecting groups for the following transformation:

OHC–(bicyclic ketone) → OHC–(bicyclic tertiary alcohol, Me, OH)

8.3. Suggest suitable protecting groups for the following transformation:

$Me_2CHCH(NH_2)CO_2H + MeCH(NH_2)CO_2H \longrightarrow Me_2CHCH(NH_2)CONHCH(Me)CO_2H$

8.4. Suggest suitable protecting groups for the following transformation:

D-glucose → erythritol ($HOCH_2CH(OH)CH(OH)CH_2OH$)

Further Reading

M. Lalonde and T. H. Chan, *Use of organosilicon reagents as protective groups in organic synthesis*, in *Synthesis*, 1985, 817.

J. Muzart, *Silyl ethers as protective groups for alcohols*, in *Synthesis*, 1993, 11.

T. D. Nelson and R. D. Crouch, *Selective deprotection of silyl ethers*, in *Synthesis*, 1996, 1031.

G. Hofle, W. Steglich and H. Vorbruggen, *4-Dialkylaminopyridines as highly active acylation catalysts*, in *Angew. Chem. Int. Ed. Engl.*, 1978, **17**, 569.

J. S. Fruchtel and G. Jung, *Organic chemistry on solid supports*, in *Angew. Chem. Int. Ed. Engl.*, 1996, **35**, 17.

G. Lowe, *Combinatorial chemistry*, in *Chem. Soc. Rev.*, 1995, **24**, 309.

S. Kobayashi, *New methodologies for the synthesis of compound libraries*, in *Chem. Soc. Rev.*, 1998, **28**, 1.

9 Some Examples of Total Synthesis

Aims

The aim of this chapter is to illustrate the dissection of target structures and the development of a synthesis. By the end of this chapter you should understand:

- The way in which a target structure is dissected to reveal a synthetic strategy
- The value of a convergent synthesis
- The way in which the relationship between functional groups reveals particular steps in a synthesis

9.1 Introduction

In the previous chapters we have considered individual synthetic steps and noted their structural outcomes. However, a synthesis involves a series of steps from the starting material to the target molecule. In this chapter we will consider the target molecule and the ways in which it may be dissected to reveal a suitable synthetic scheme. In doing this it is worth remembering that there can be a number of different solutions to the same synthetic problem.

Four target molecules have been chosen to exemplify strategies using Wittig, Grignard and ring extension reactions (β-eudesmol), aromatic substitution and Michael addition reactions (griseofulvin), heterocyclic synthesis (vitamin B_1) and organometallic methods (prostaglandin E_2). The synthesis of β-eudesmol is a linear synthesis, but the syntheses of griseofulvin, vitamin B_1 and prostaglandin E_2 are convergent. The syntheses of β-eudesmol and griseofulvin produced racemic material, but that of the prostaglandin E_2 has an optically active chiral starting material.

9.2 β-Eudesmol

β-Eudesmol is a widespread sesquiterpene alcohol. It is a *trans*-decalin with an equatorial propan-2-ol substituent and an exocyclic double bond which undergoes isomerization into the ring to form the α- and γ-isomers. A synthesis[1] of this compound reveals the general strategy of removing the appendages to identify the underlying carbon skeleton. A retrosynthetic analysis is shown in Scheme 9.1. The regiospecific construction of the exocyclic 4(15)-methylene implies the use of a Wittig reaction on the corresponding ketone. The tertiary alcohol in the side chain is typical of the product of a Grignard reaction on an ester. This ester in turn could be constructed by the carboxylation of a Grignard reagent derived from a halogen at C-7. Since two potentially competing carbonyl addition reactions, a Wittig and a Grignard sequence, are considered, a carbonyl group cannot be introduced at C-4 until the Grignard sequence is complete. The synthesis therefore rested on the construction of a decalin with functional groups at C-4 and C-7 which could be transformed into a ketone and a halogen, respectively. There are well-established routes for the construction of decalins based, for example, on the Robinson ring-extension reaction. This dissection is shown in Scheme 9.2.

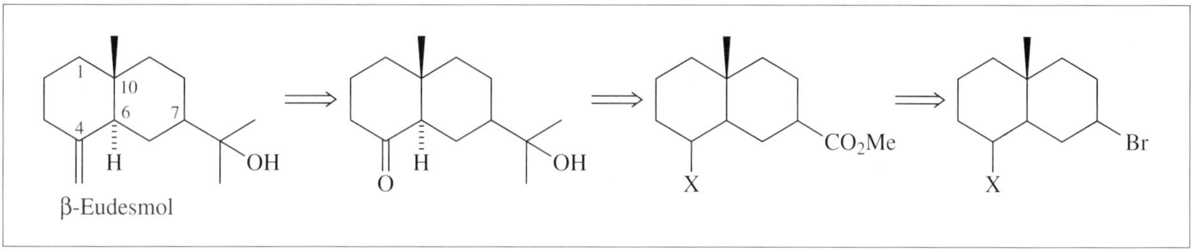

Scheme 9.1

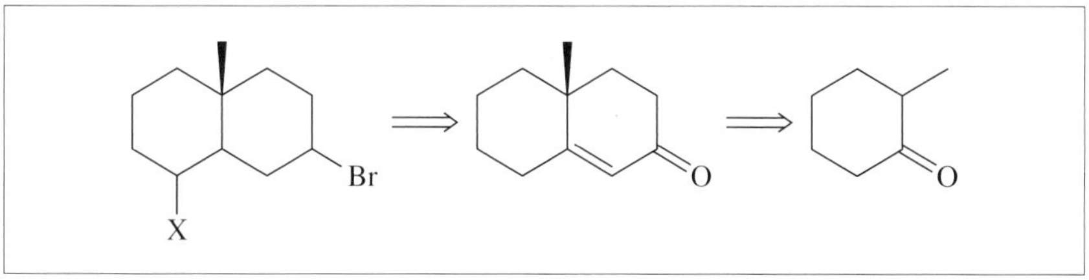

Scheme 9.2

The synthesis of racemic β-eudesmol used the following sequence of reactions (Scheme 9.3). The bicyclic carbon skeleton was prepared by a Robinson ring-extension reaction from 2-methylcyclohexanone. The unsaturated ketone was then converted to its dienol benzoate, which was reduced with sodium borohydride to form the unsaturated alcohol. The alcohol was transformed into the bromide by treatment with phosphorus tribromide. Interestingly, there was retention of configuration in this reaction due to the participation of the double bond in forming a homoallylic carbocation. The bromide was converted to its Grignard derivative, and this was reacted with carbon dioxide to form the acid and thence, with diazomethane, the methyl ester. The ester was stable to prolonged treatment with sodium methoxide and hence it was an equatorial substituent. Treatment of the ester with methylmagnesium iodide gave the tertiary alcohol. The alkene was then converted to the 4-ketone by hydroboration and oxidation. The stereochemistry of the hydroboration of the alkene was only partially determined by the angular methyl group at C-10, and both isomers at C-5 were obtained. However, equilibration of the 4-ketone with sodium methoxide allowed enolization to take place and converted all the material to the more stable *trans*-decalone. The synthesis was completed by a Wittig reaction.

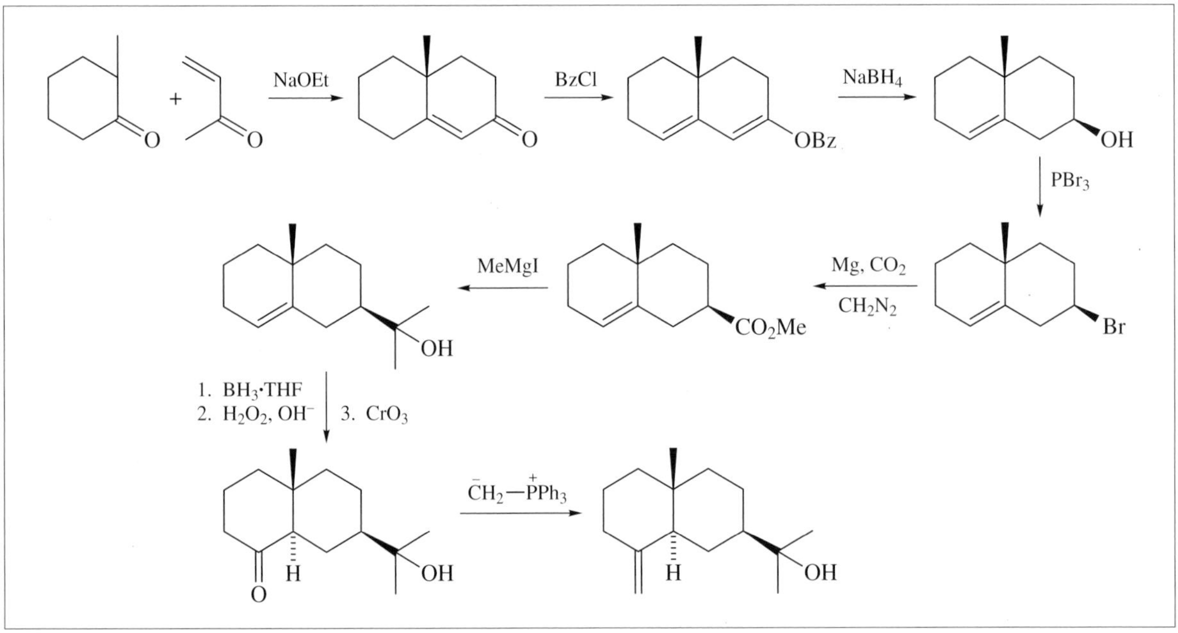

Scheme 9.3

The synthesis reveals not only the simplification of the synthetic problem by removing the appendages, but also the use of equilibration methods to obtain and identify the more stable isomers.

9.3 Griseofulvin

The antibiotic griseofulvin has been used in the treatment of fungal infections of the skin. It has been obtained from a number of fungi, including *Penicillium griseofulvum*. The study of its biosynthesis has played an important part in providing experimental evidence for polyketide biosynthesis and the role of phenol coupling in biosynthesis.

Retrosynthetic analysis (Scheme 9.4) of the structure of griseofulvin showed that there was a 1,5-relationship between the carbonyl groups, which suggested that Michael reactions could play an important part in a synthesis. The dissection exposes two fragments of quite different reactivity: a coumaranone and a methoxyethynyl propenyl ketone. This analysis leads to a convergent synthesis.[2]

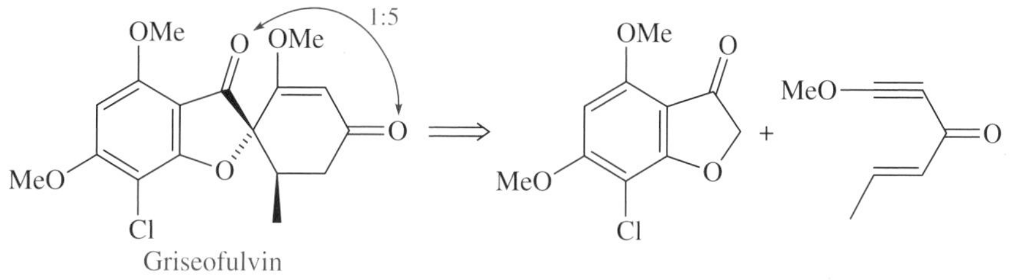

Scheme 9.4

The benzenoid ring of the coumaranone (Scheme 9.5) contains a symmetrical 1,3,5-trihydroxybenzene (a phloroglucinol unit) with a halogen and a deactivating carbonyl attached to it. The order of introduction of substituents onto an aromatic ring can be important, with deactivating substituents being introduced later in a synthesis. The formation of the oxygen ring could involve the displacement of chlorine from a chloroacetophenone.

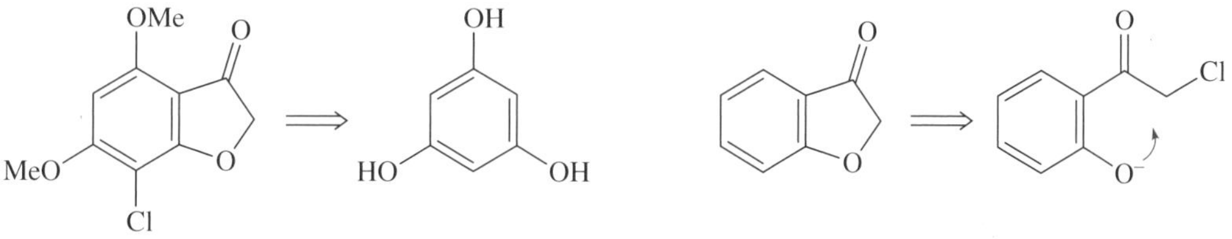

Scheme 9.5

The synthesis of this part of the molecule was achieved as follows (Scheme 9.6). 1,3,5-Trihydroxybenzene was converted to its dimethyl ether and chlorinated to give a separable mixture of 2-chloro-3,5-dimethoxyphenol and 4-chloro-3,5-dimethoxyphenol. Friedel–Crafts acylation of the former with chloroacetyl chloride gave the chloroacetophenone, which underwent cyclization in the presence of sodium acetate to form the coumaranone.

Retrosynthetic analysis (Scheme 9.7) of the alkynyl ketone suggested that it might be obtained via the alkynyl alcohol, which would arise from the addition of methoxyethyne to *trans*-but-2-enal (crotonaldehyde). An alkynyl unit is rarely synthesized as such within a molecule but is more often inserted into a molecule.

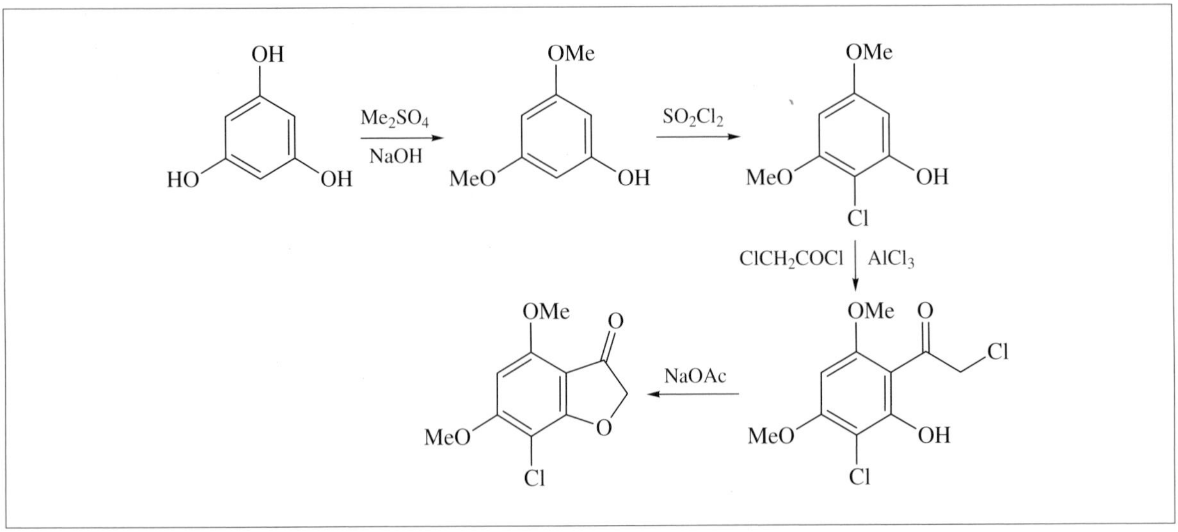

OH
HO OH
Me_2SO_4
NaOH
OMe
MeO OH
SO_2Cl_2
OMe
MeO OH
Cl
$ClCH_2COCl$ | $AlCl_3$
OMe O
Cl
MeO OH
Cl
NaOAc
OMe O
MeO O
Cl

Scheme 9.6

MeO—≡ =O ⟹ MeO—≡ H OH ⟹ MeO—≡ + =O

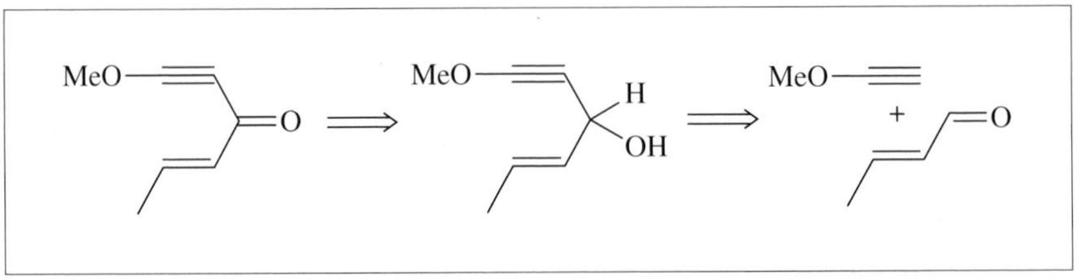

Scheme 9.7

The synthesis (Scheme 9.8) involved the condensation of the lithium salt of methoxyethyne with *trans*-but-2-enal to give the alkynyl alcohol, which was oxidized with manganese dioxide to form the relatively unstable alkynyl propenyl ketone. The two portions of the molecule were then linked together in a double Michael condensation in the presence of potassium *t*-butoxide (Scheme 9.8). Epimers of griseofulvin are possible at both the spiro centre and on the cyclohexenone ring (the C-2 methyl group). Fortunately this synthesis led to the correct stereoisomer.

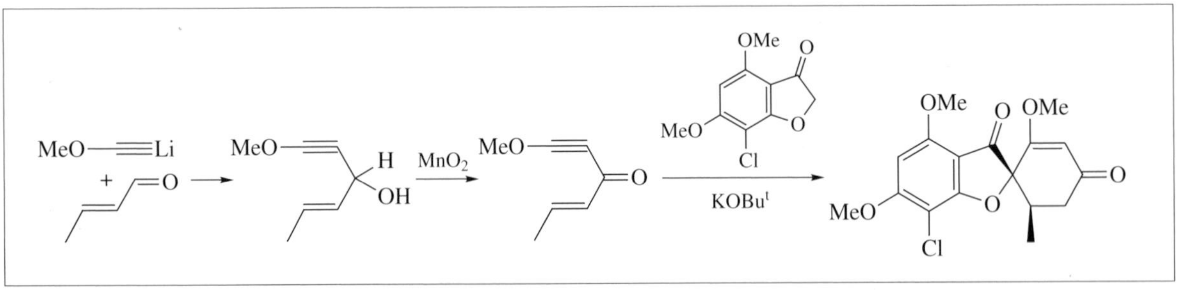

MeO—≡Li + =O → MeO—≡ H OH MnO_2 → MeO—≡ =O
OMe O
MeO O
Cl
$KOBu^t$
OMe O OMe
MeO O =O
Cl

Scheme 9.8

The biogenetic speculation on the role of phenol coupling in the formation of griseofulvin predicted a key role for griseophenone A. Hence a biogenetically patterned synthesis (Scheme 9.9) was developed.[3] The benzophenone griseophenone A was prepared by a Friedel–Crafts reaction between 2-chloro-3,5-dimethoxyphenol and the appropriate benzoyl chloride. When this was oxidized under mild conditions with potassium hexacyanoferrate(III), the dienone dehydrogriseofulvin was obtained. Controlled reduction of this gave griseofulvin. Biogenetically patterned syntheses can shed a very interesting light on the chemistry of enzymatic processes.

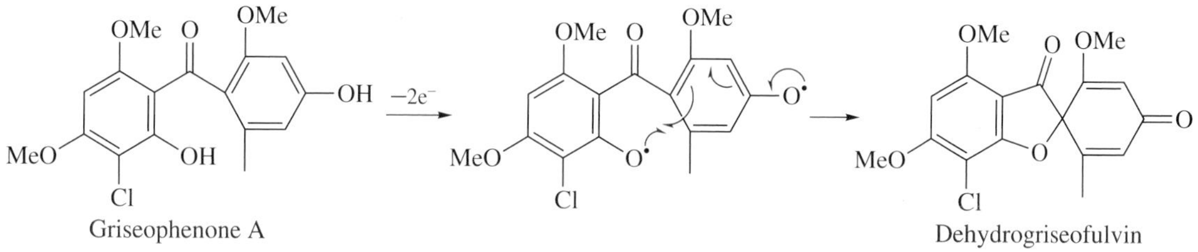

Griseophenone A $\xrightarrow{-2e^-}$ → Dehydrogriseofulvin

Scheme 9.9

9.4 Thiamine (Vitamin B_1)

Thiamine (vitamin B_1 or aneurin) is one of the original essential dietary factors for which the term vitamin was coined in 1911. A deficiency of vitamin B_1 leads to the disease beri-beri. Thiamine, as its pyrophosphate, is a co-factor in a number of important metabolic processes.

Retrosynthetic analysis of the structure (Scheme 9.10) shows that it contains two heterocyclic rings, a pyrimidine and a thiazole, linked by a methylene attached to a quaternary nitrogen. The dissection into two halves at this point represents a reasonable first retrosynthetic step, and led to a convergent synthesis.[4] Note that the halogen (group X in Scheme 9.10), attached to the carbon which would be used to alkylate the nitrogen of the thiazole ring, is an allylic and hence reactive halogen.

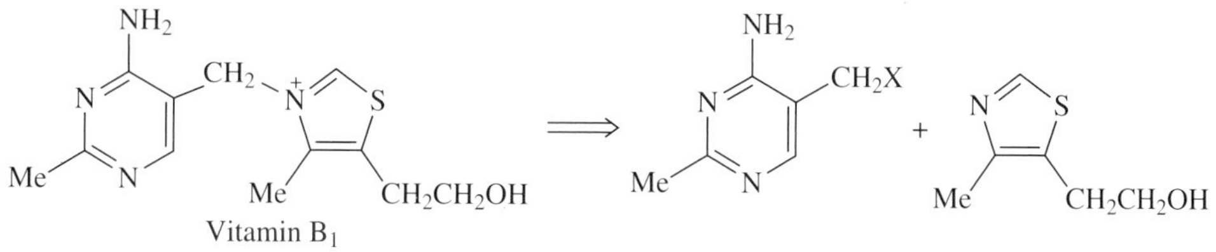

Vitamin B_1

Scheme 9.10

The retrosynthetic analysis of the heterocyclic rings may be helped by mentally replacing the nitrogen atoms by oxygen to see if this reveals any potentially useful carbonyl–amine condensation reactions. There are well-established pyrimidine syntheses based on the condensation of ureas and their derivatives with 1,3-dicarbonyl compounds which can be identified this way. Carrying out this dissection (Scheme 9.11) on the pyrimidine suggests a condensation between acetamidine and a dicarbonyl compound. Analysis of the carbonyl component shows that the oxidation levels of the terminal carbons correspond to a carboxylic acid (amide), an aldehyde and a primary alcohol (ether). The synthesis of the pyrimidine component used the following route (Scheme 9.12). Ethyl β-ethoxypropanoate was condensed with ethyl formate in the presence of sodium and then treated with acetamidine hydrochloride in the presence of sodium ethoxide to give 5-(ethoxymethyl)-6-hydroxy-2-methylpyrimidine. Chlorination with phosphorus oxychloride and amination with alcoholic ammonia gave the aminopyrimidine. Note that this displacement involves the formation of an imino chloride [–N=C(R)–Cl], which bears a formal similarity to an acyl chloride [O=C(R)–Cl] and has comparable reactivity.

Scheme 9.11

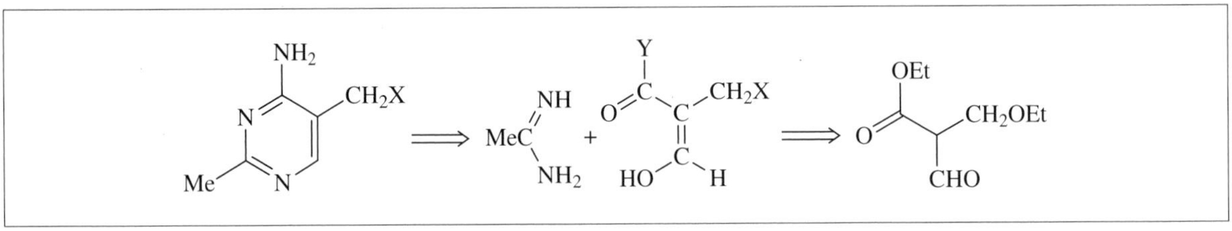

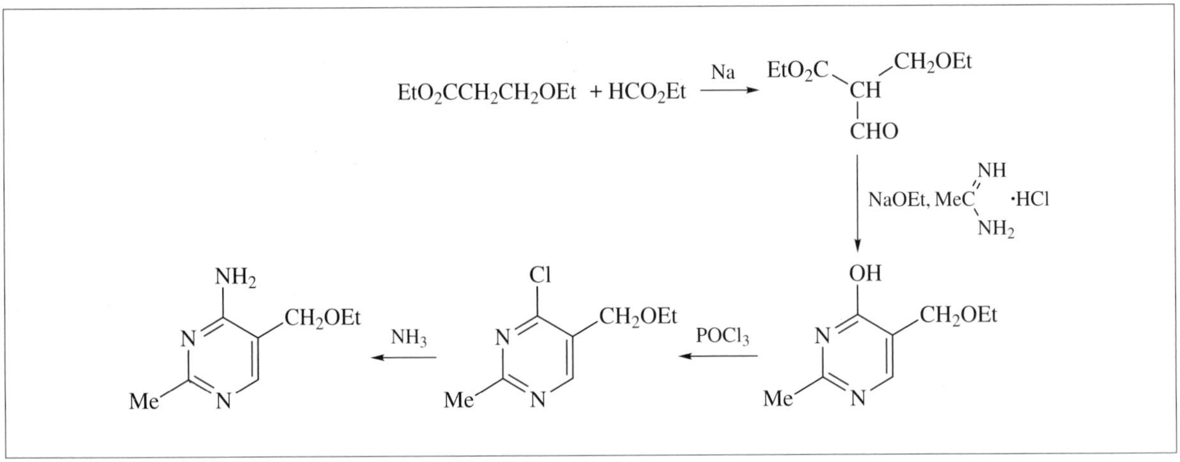

Scheme 9.12

Turning to the thiazole portion, and carrying out a similar replacement of the nitrogen by oxygen, suggests the following dissection (Scheme 9.13) based on thioformamide.

The preparation of the halogenated pentanone presents two problems. The halogen must be introduced regiospecifically and in the presence of a primary alcohol. The latter obviously needs protecting, for example as the acetate, but the requirement for regiospecificity requires an activation of a particular centre. The clue (Scheme 9.14) to this comes from the

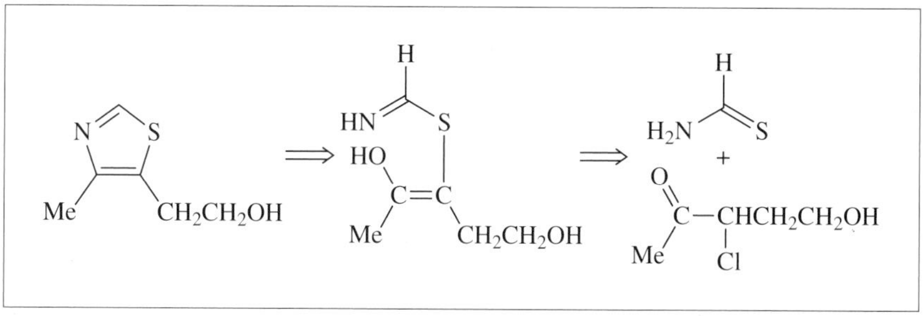

$$\text{thiazole (N, S; Me, } CH_2CH_2OH) \Longrightarrow HN{=}CH{-}S{-}C(CH_2CH_2OH){=}C(Me)OH \Longrightarrow H_2NCH{=}S + MeC(=O){-}CH(Cl)CH_2CH_2OH$$

Scheme 9.13

presence in the pentanone of a MeC(R)=O group which is a "marker" for the use of ethyl acetoacetate.

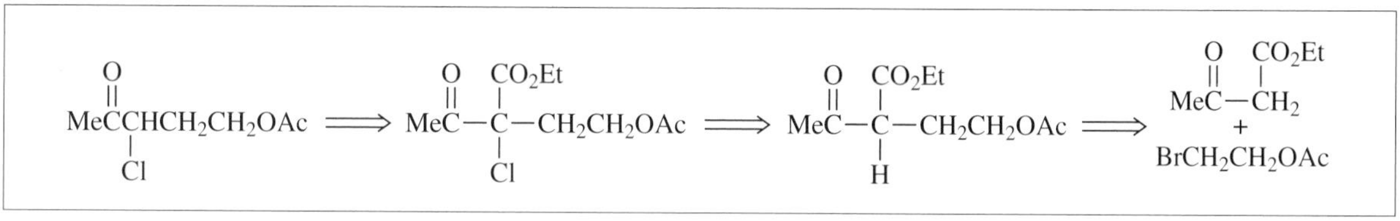

$$MeC(=O)CH(Cl)CH_2CH_2OAc \Longrightarrow MeC(=O){-}C(Cl)(CO_2Et){-}CH_2CH_2OAc \Longrightarrow MeC(=O){-}CH(CO_2Et){-}CH_2CH_2OAc \Longrightarrow MeC(=O){-}CH_2CO_2Et + BrCH_2CH_2OAc$$

Scheme 9.14

The synthesis (Scheme 9.15) of the thiazole portion involved alkylation of ethyl acetoacetate with 1-acetoxy-2-bromoethane, and then chlorination with sulfuryl chloride and careful acid-catalysed hydrolysis and decarboxylation.

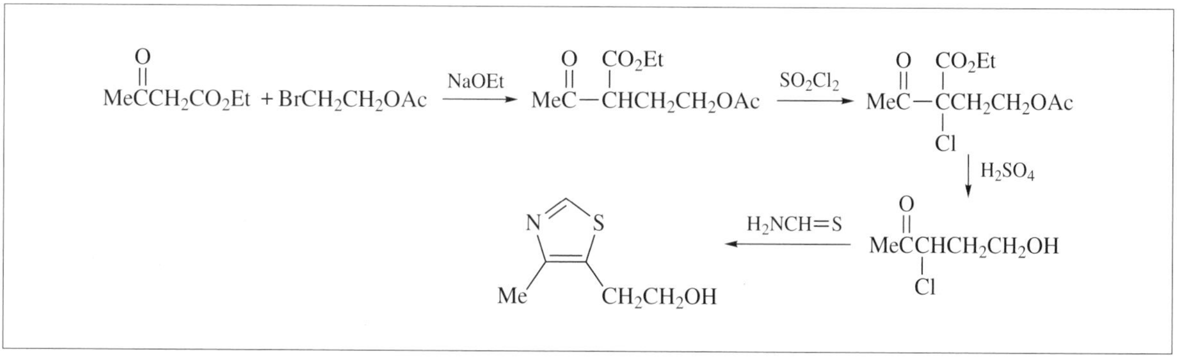

$$MeC(=O)CH_2CO_2Et + BrCH_2CH_2OAc \xrightarrow{NaOEt} MeC(=O){-}CH(CO_2Et)CH_2CH_2OAc \xrightarrow{SO_2Cl_2} MeC(=O){-}C(Cl)(CO_2Et)CH_2CH_2OAc \xrightarrow{H_2SO_4} MeC(=O)CH(Cl)CH_2CH_2OH \xrightarrow{H_2NCH{=}S} \text{thiazole (Me, } CH_2CH_2OH)$$

Scheme 9.15

In order to link the two portions together, the allylic ether was cleaved with hydrogen bromide to give a reactive allylic bromide. At the same time, the potentially nucleophilic amino group on the pyrimidine ring was converted to its hydrobromide salt, and hence dimerization did not occur. The vitamin was then prepared (Scheme 9.16) by displacing the allylic bromide with the thiazole. This particular synthesis has been carried out commercially on a large scale.

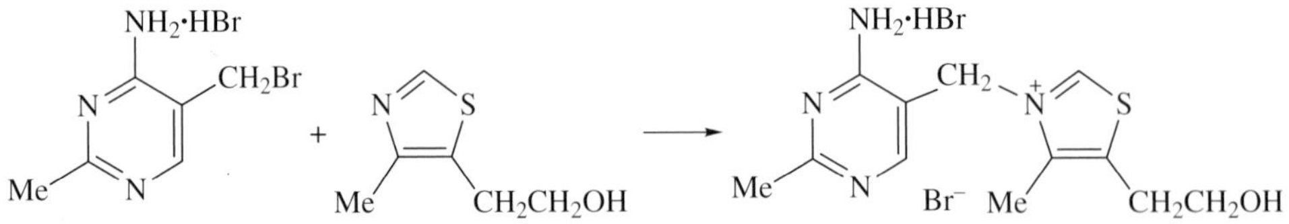

Scheme 9.16

9.5 Prostaglandins

The prostaglandins (Scheme 9.17) are a widespread group of hormones which control a range of biological processes. They are present in very small amounts in biological material and hence studies of their biological effects have relied on material which has been made by total synthesis. The synthetic problems involve constructing three (PGE series) or four (PGF series) adjacent chiral centres around a five-membered ring (see Scheme 9.17). In the side chains there are two double bonds, one with a *cis* and the other with a *trans* geometry, and there is also a chiral alcohol in one side chain.

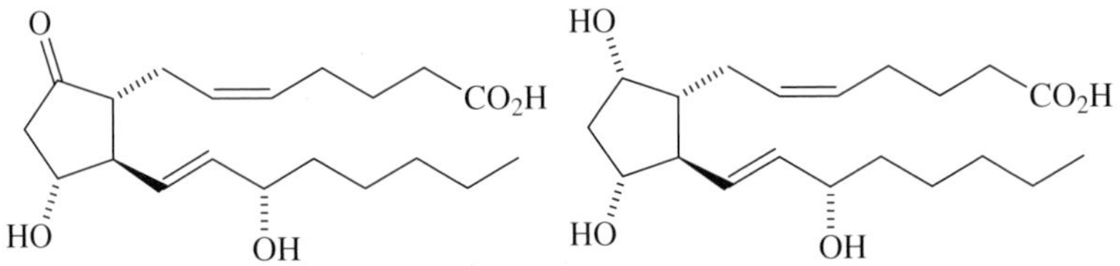

Prostaglandin E_2 Prostaglandin F_2

Scheme 9.17

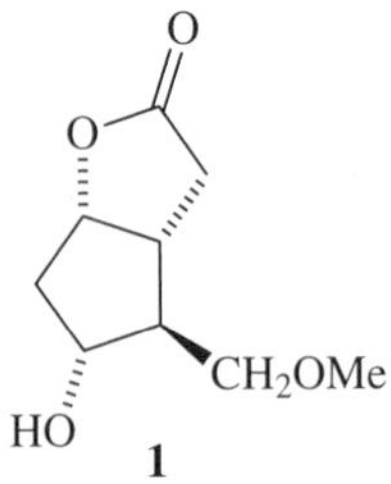

There are very many solutions to these synthetic problems, some of which have been used on a large scale. Many of the strategies have concentrated on constructing the centres around the five-membered ring, and then adding the side chains using the double bonds as the linking points. The total synthesis by Corey[5] in 1969 made use of cyclopentadiene to form the lactone **1** containing all four asymmetric centres, the addition of the side chains using a Wadsworth–Emmons phosphonate procedure to create the *trans* double bond, and a Wittig procedure to create the *cis* double bond.

A simple strategy[6] for the synthesis of prostaglandin E_2 made use of organometallic chemistry. Two dissections (Scheme 9.18) based on the carbonyl group are possible. The adjacent position can be alkylated via an enolate to attach the carboxyl side chain, and the enolate itself could be generated by the addition of a cuprate to the β-position of a cyclopentenone. The stereochemistry of these additions would be determined by the stereochemistry of the group in the γ-position. Although this three-component disconnection is conceptually elegant, a great deal of work had to go on in order to bring the synthesis (Scheme 9.19) to fruition.

Scheme 9.18

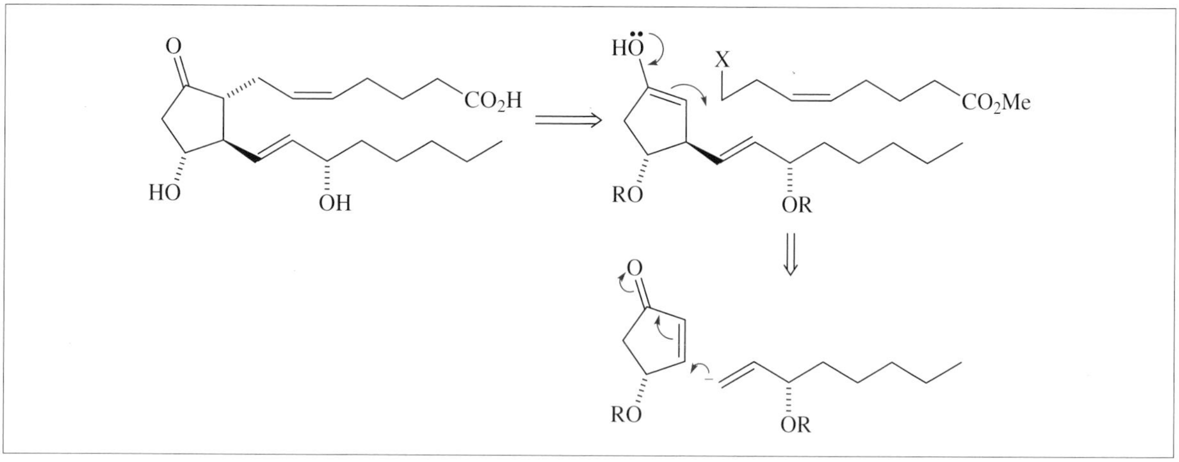

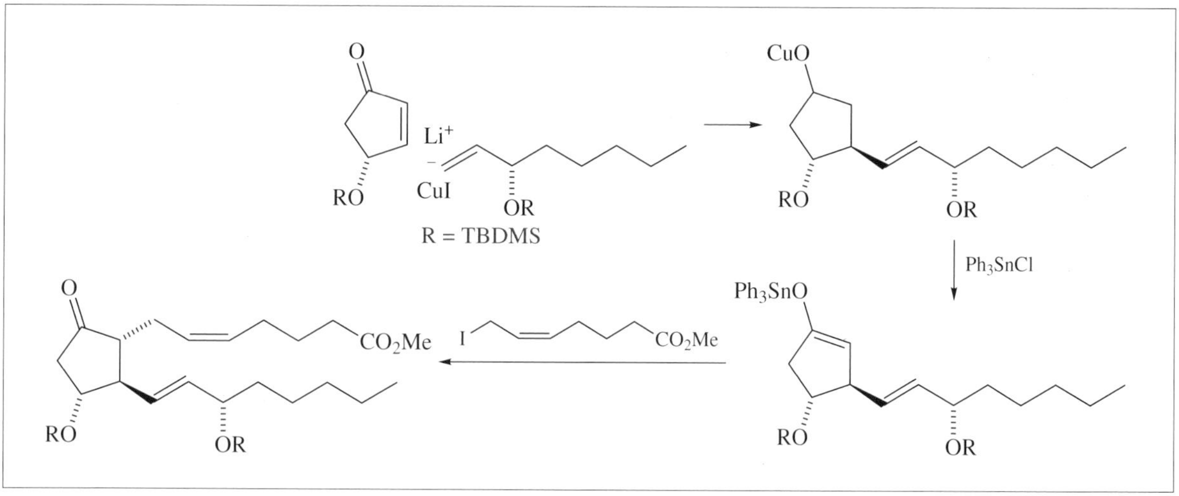

Scheme 9.19

The optically active hydroxycyclopentenone (Scheme 9.19) was obtained from cyclopentadiene using both chemical and microbiological methods to obtain chiral material. Addition of an organocopper reagent to the enone gave an enolate. The stereochemistry of the addition was directed by the bulky protecting group R on the γ-hydroxyl. The copper enolate was then converted to a tin enolate, which was alkylated with an allylic iodide to give a PGE_2 derivative. This synthesis reflects the regio- and stereochemical control over synthetic reactions provided by organometallic reagents.

References

1. C. H. Heathcock and T. Ross Kelly, *Tetrahedron*, 1968, **24**, 1801.
2. G. Stork and M. Tomasz, *J. Am. Chem. Soc.*, 1962, **84**, 310.
3. A. C. Day, J. Nabney and A. I. Scott, *J. Chem. Soc.*, 1961, 4067.
4. J. K. Cline, R. R. Williams and J. Finkelstein, *J. Am. Chem. Soc.*, 1937, **59**, 1052.
5. E. J. Corey, N. M. Weinshenker, T. K. Schaaf and W. Huber, *J. Am. Chem. Soc.*, 1969, **91**, 5675.
6. M. Suzuki, A. Yanagisawa and R. Noyori, *J. Am. Chem. Soc.*, 1985, **107**, 3348.

Further Reading

The study of organic synthesis relies on a firm grasp of functional group chemistry, organic reaction mechanisms and stereochemistry, together with the more detailed aspects of aromatic and heterocyclic chemistry. These topics are covered in further books in this series.[1]

Many of the reactions that are used in organic synthesis are amplified in the larger textbooks of organic chemistry.[2] More detailed discussions of the relevant chemistry of functional groups and synthetic methods can be found in the multi-volume treatises such as *Rodd's Chemistry of Carbon Compounds*,[3] *Comprehensive Organic Chemistry*,[4] *Comprehensive Organic Synthesis*,[5] and *Comprehensive Functional Group Transformations*.[6] The chemistry of the individual functional groups is described in detail in the series by Patai.[7]

A useful book to turn to for references to the use of particular reactions is *Comprehensive Organic Transformations* by Larock.[8] More detailed discussions of the mechanisms of these reactions are given in *Organic Synthesis* by Smith[9] and March's *Advanced Organic Chemistry*.[10] These books also give many references to the original literature. The older literature describing fundamental studies on the chemistry of alkenes, aromatic compounds and carbonyl compounds is brought together and referenced in Royals' *Advanced Organic Chemistry*.[11] Many references and brief descriptions of older methods are also to be found in Hickinbottom's *Reactions of Organic Compounds*.[12]

The scope and mechanism of many of the named reactions that are used in synthetic organic chemistry are thoroughly reviewed in the series *Organic Reactions*.[13] Specific examples of synthetic procedures are collected together in *Organic Syntheses*[14] and in Vogel's *Textbook of Practical Organic Chemistry*.[15] Regular surveys of organic synthetic methods are published in Theilheimer's *Synthetic Methods of Organic Chemistry*,[16] Houben-Weyl's *Stereoselective Synthesis*[17] and in *A Compendium of Organic Synthetic Methods*.[18] There are a number of series such as the *Best Synthetic Methods* series[19] and the *Postgraduate*

Chemistry Series[20] which contain volumes devoted to particular types of transformation. Advances in the total synthesis of groups of natural products are brought together in the series *The Total Synthesis of Natural Products*.[21]

Detailed information on the reagents that are used in the transformations of functional groups is collected together in Fieser's *Reagents for Organic Synthesis*[22] and in the *Encyclopedia of Reagents for Organic Synthesis.*[23] The information in these volumes has also been brought together in a series of handbooks of reagents for organic synthesis. There are several books describing the use of protecting groups in organic synthesis.[24]

There are a number of books that describe the strategies of organic synthesis. These include Corey's *The Logic of Chemical Synthesis*.[25] The disconnection approach to organic synthesis is explored in more detail in *Organic Synthesis, The Disconnection Approach.*[26]

Recent advances in the chemistry of functional groups in a synthetic context were reviewed in the journal *Contemporary Organic Synthesis* and are now covered in review articles which appear in the *Journal of the Chemical Society, Perkin Transactions 1*. More specific reviews appear in other journals such as *Synthesis*, *Synlett*, *Angewandte Chemie (International Edition)*, *Tetrahedron* and *Aldrichimica Acta*. Other useful reviews have appeared in *Chemical Society Reviews* and its predecessor, *Quarterly Reviews of the Chemical Society*. The American Chemical Society equivalent is *Chemical Reviews*. There are a number of journals which contain descriptions of new methodology such as the *Journal of Organic Chemistry*, *Synthetic Communications* and the *Journal of Chemical Research*.

References

1. (a) D. G. Morris, *Stereochemistry*, RSC Tutorial Chemistry Series, Royal Society of Chemistry, Cambridge, 2001; (b) J. R. Hanson, *Functional Group Chemistry*, Royal Society of Chemistry, Cambridge, 2001.
2. See, for example, J. Claydon, N. Greaves, S. Warren and P. Wothers, *Organic Chemistry*, Oxford University Press, Oxford, 2000; M. Jones, *Organic Chemistry*, 2nd edn., Norton, New York, 2000; T. W. G. Solomons and C. Fryhle, *Organic Chemistry*, 7th edn., Wiley, New York, 2000; K. P. C. Volhardt and N. E. Schore, *Organic Chemistry*, 3rd edn., Freeman, New York, 1998; R. J. Fessenden, J. S. Fessenden and M. Logue, *Organic Chemistry*, 6th edn., Brooks/Cole, Pacific Grove,

Calif., 1998; R. O. C. Norman and J. Coxon, *Principles of Organic Synthesis*, 3rd edn., Blackie, Glasgow, 1993.

3. *Rodd's Chemistry of Carbon Compounds*, 2nd edn., ed. S. Coffey, Elsevier, Amsterdam, 1964 et seq.
4. *Comprehensive Organic Chemistry*, ed. D. H. R. Barton and W. D. Ollis, Pergamon Press, Oxford, 1979.
5. *Comprehensive Organic Synthesis*, ed. B. M. Trost and I. Fleming, Pergamon Press, Oxford, 1991.
6. *Comprehensive Functional Group Transformations*, ed. A. R. Katritzky, O. Meth-Cohn and C. W. Rees, Pergamon Press, Oxford, 1995.
7. *The Chemistry of Functional Groups*, ed. S. Patai, Wiley, Chichester, 1964–present.
8. R. C. Larock, *Comprehensive Organic Trasnformations*, 2nd edn., VCH, New York, 1999.
9. M. B. Smith, *Organic Synthesis*, McGraw-Hill, New York, 1994.
10. J. March, *Advanced Organic Chemistry*, 4th edn., Wiley, New York, 1992.
11. E. Royals, *Advanced Organic Chemistry*, Constable, London, 1954.
12. W. J. Hickinbottom, *Reactions of Organic Compounds*, 3rd edn., Longmans, London, 1957.
13. *Organic Reactions*, vols. 1–53, Wiley, New York, 1942–1998.
14. *Organic Syntheses*, vols. 1–76, Wiley, New York, 1932–present; also *Organic Syntheses Collective Volumes*, 1–8.
15. A. R. Tatchell et al., *Vogel's Textbook of Practical Organic Chemistry*, 5th edn., Longmans, Harlow, 1989.
16. *Theilheimer's Synthetic Methods of Organic Chemistry*, ed. A. F. Finch, Karger/Derwent, Basle/London, vols. 1–57, 1948–1999.
17. *Houben-Weyl's Stereoselective Synthesis*, vols. 1–10, ed. G. Helmchen, Thieme, Stuttgart, 1996.
18. *A Compendium of Organic Synthetic Methods*, vols. 1–8, Wiley, New York, 1971–1995.
19. *Best Synthetic Methods*, Academic Press, San Diego.
20. *Postgraduate Chemistry Series*, Sheffield Academic Press, Sheffield.
21. *The Total Synthesis of Natural Products*, vols. 1–11, ed. J. W. Apsimon, Wiley, New York, 1973–1999.
22. L. Fieser and M. Fieser, *Reagents for Organic Synthesis*, vols. 1–19, Wiley, New York, 1968–1999.

23. *Encyclopedia of Reagents for Organic Synthesis*, vols. 1–8, ed. L. A. Paquette, Wiley, Chichester, 1995; *Handbook of Reagents for Organic Synthesis*, vols. 1–4, Wiley, Chichester, 1999.

24. P. J. Kocienski, *Protecting Groups*, Thieme, New York, 1994; T. W. Greene and P. G. M. Wuts, *Protective Groups in Organic Synthesis*, 3rd edn., Wiley, New York, 1999; J. R. Hanson, *Protecting Groups in Organic Synthesis*, Sheffield Academic Press, Sheffield, 1999.

25. E. J. Corey and X. M. Cheng, *The Logic of Chemical Synthesis*, Wiley, New York, 1995.

26. S. Warren, *Organic Synthesis, The Disconnection Approach*, Wiley, Chichester, 1982.

Answers to Problems

Chapter 1

1.1.

(a) TM $\Longrightarrow$ HC≡CNa + $Me_2C{=}O$

TM = target molecule

(b) 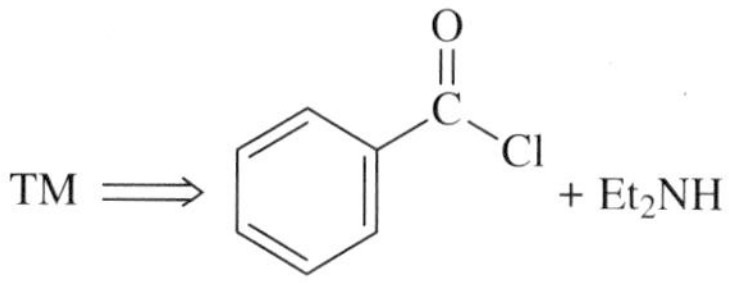 TM $\Longrightarrow$ PhCOCl + Et_2NH

(c) 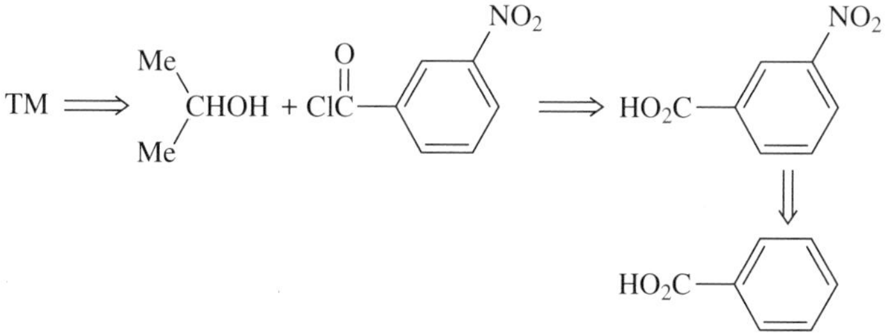 TM $\Longrightarrow$ Me_2CHOH + ClCO–C_6H_4–NO_2 (*m*) $\Longrightarrow$ HO_2C–C_6H_4–NO_2 (*m*) $\Longrightarrow$ HO_2C–C_6H_5

(d) 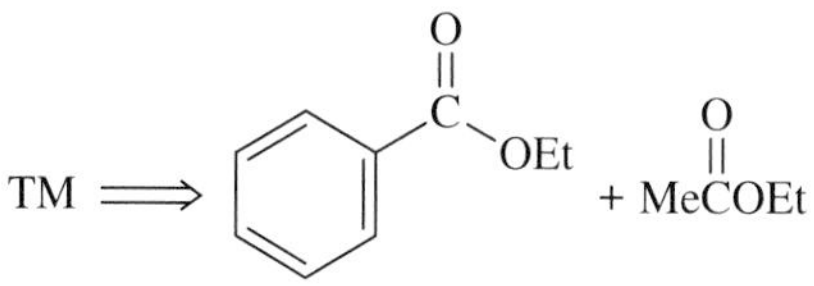 TM $\Longrightarrow$ PhCOOEt + MeCOOEt

(e) 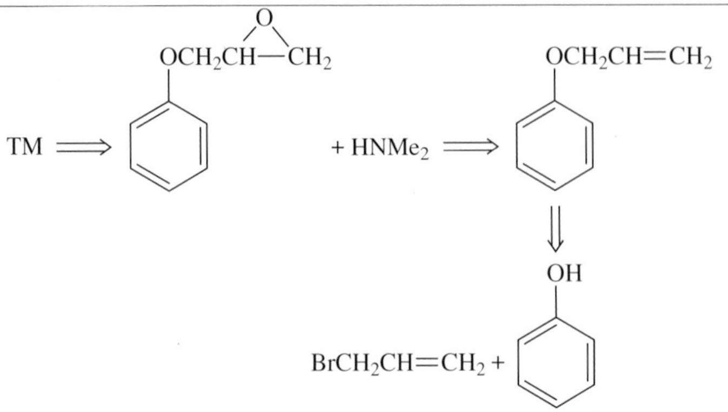 TM $\Longrightarrow$ $PhOCH_2CH(O)CH_2$ (epoxide) + $HNMe_2$ $\Longrightarrow$ $PhOCH_2CH{=}CH_2$

$\Downarrow$

$BrCH_2CH{=}CH_2$ + PhOH

(f) 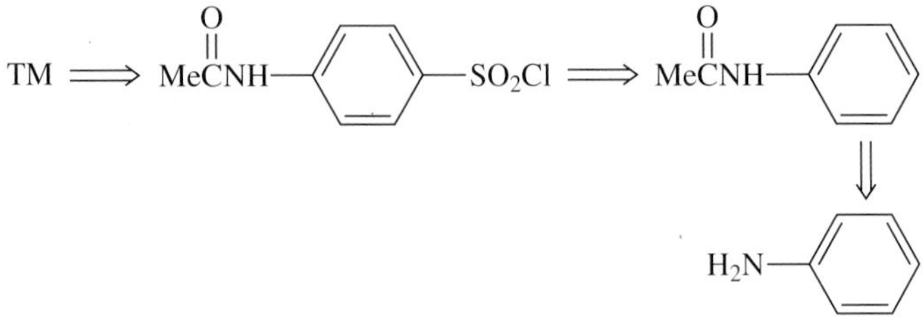 TM $\Longrightarrow$ $MeC(=O)NH{-}C_6H_4{-}SO_2Cl$ $\Longrightarrow$ $MeC(=O)NH{-}C_6H_5$

$\Downarrow$

$H_2N{-}C_6H_5$

Chapter 2

2.1.

(a) $^{14}CH_3I \xrightarrow[Et_2O]{Mg} {}^{14}CH_3MgI \xrightarrow[2.\ H^+]{1.\ CO_2} {}^{14}CH_3CO_2H$

(b) $MeI \xrightarrow[Et_2O]{Mg} MeMgI \xrightarrow[2.\ H^+]{1.\ ^{14}CO_2} Me^{14}CO_2H$

(c) $^{14}CH_3I \xrightarrow[Et_2O]{Mg} {}^{14}CH_3MgI \xrightarrow[2.\ H^+]{1.\ CH_2{-}CH_2\ (\text{ethylene oxide})} {}^{14}CH_3CH_2CH_2OH$

(d) $MeI \xrightarrow[Et_2O]{Mg} MeMgI \xrightarrow[H^+]{H^{14}CHO} Me^{14}CH_2OH \xrightarrow{PCl_5} Me^{14}CH_2Cl$

$Me^{14}CH_2Cl \xrightarrow{Mg,\ Et_2O} Me^{14}CH_2MgCl \xrightarrow[H^+]{HCHO} Me^{14}CH_2CH_2OH$

(e) $EtI \xrightarrow[Et_2O]{Mg} EtMgI \xrightarrow[H^+]{H^{14}CHO} Et^{14}CH_2OH$

(f) $Me^{14}CO_2H$ (from b) $\xrightarrow{SOCl_2} Me^{14}C(=O)Cl \xrightarrow{Et_2Cd} Me^{14}C(=O)Et$

2.2.

(a) $MeCH{=}CHC(=O)Me \xrightarrow[CuI]{^{13}CH_3MgI} (Me)(^{13}CH_3)CHCH_2C(=O)Me$

(b) 2-cyclohexen-1-one (O) $+ MeC(MgBr){=}CH_2 \xrightarrow{CuI}$ 3-isopropenylcyclohexanone (O)

2.3.

(a) $HC{\equiv}CNa + MeC(=O)CH_2Me \longrightarrow HC{\equiv}CC(Me)(OH){-}CH_2Me$

(b) $BuBr + NaC{\equiv}CH \longrightarrow BuC{\equiv}CH \xrightarrow{MeMgBr} BuC{\equiv}CMgBr$

$BuC{\equiv}CMgBr \xrightarrow{MeCHO} BuC{\equiv}CCH(OH)Me$

$BuC{\equiv}CCH(OH)Me \xrightarrow{CrO_3} BuC{\equiv}CC(=O)Me$

2.4.

$CH_2{-}CH_2$ (with bridging O) $+ H^{13}C{\equiv}^{13}CH \xrightarrow{NaNH_2} HOCH_2CH_2{}^{13}C{\equiv}^{13}CH$

$HOCH_2CH_2{}^{13}C{\equiv}^{13}CH \xrightarrow{MeMgI} BrMgOCH_2CH_2{}^{13}C{\equiv}^{13}CMgBr$

$BrMgOCH_2CH_2{}^{13}C{\equiv}^{13}CMgBr \xrightarrow{CO_2} HOCH_2CH_2{}^{13}C{\equiv}^{13}CCO_2H$

$HOCH_2CH_2{}^{13}C{\equiv}^{13}CCO_2H \xrightarrow{H_2,\ Pd/BaCO_3}$ dihydropyranone (O, O; 13, 13)

2.5.

cyclohexanone (O) $+ (EtO)_2P(=O)CH_2CO_2Et \xrightarrow{NaH}$ cyclohexylidene $=CHCO_2Et$

2.6.

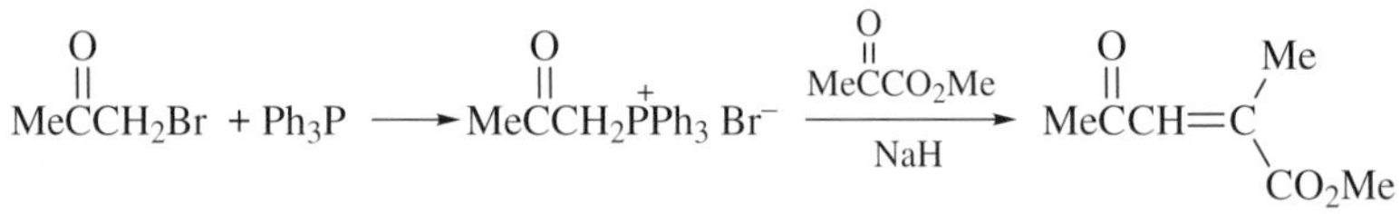

$$\mathrm{MeC(=O)CH_2Br} + \mathrm{Ph_3P} \longrightarrow \mathrm{MeC(=O)CH_2\overset{+}{P}Ph_3\ Br^-} \xrightarrow[\text{NaH}]{\mathrm{MeC(=O)CO_2Me}} \mathrm{MeC(=O)CH{=}C(Me)CO_2Me}$$

2.7.

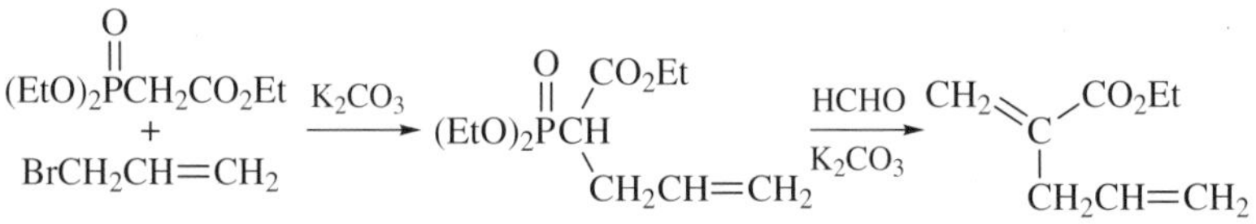

$$\mathrm{(EtO)_2P(=O)CH_2CO_2Et} + \mathrm{BrCH_2CH{=}CH_2} \xrightarrow{K_2CO_3} \mathrm{(EtO)_2P(=O)CH(CO_2Et)CH_2CH{=}CH_2} \xrightarrow[K_2CO_3]{\text{HCHO}} \mathrm{CH_2{=}C(CO_2Et)CH_2CH{=}CH_2}$$

2.8.

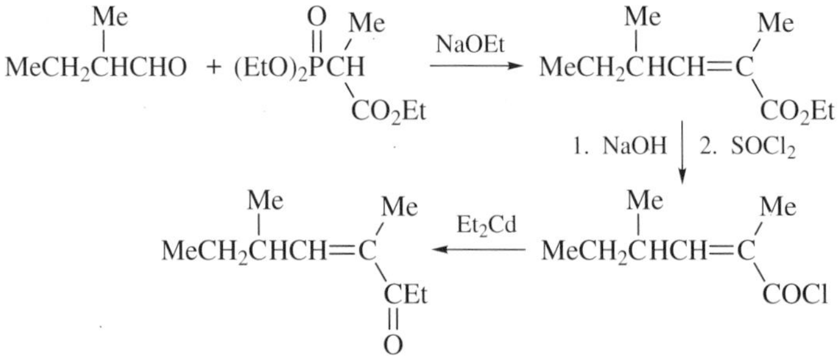

$$\mathrm{MeCH_2CH(Me)CHO} + \mathrm{(EtO)_2P(=O)CH(Me)CO_2Et} \xrightarrow{\text{NaOEt}} \mathrm{MeCH_2CH(Me)CH{=}C(Me)CO_2Et}$$

$$\xrightarrow{\text{1. NaOH} \quad \text{2. } SOCl_2} \mathrm{MeCH_2CH(Me)CH{=}C(Me)COCl} \xrightarrow{Et_2Cd} \mathrm{MeCH_2CH(Me)CH{=}C(Me)C(=O)Et}$$

Chapter 3

3.1.

(a)

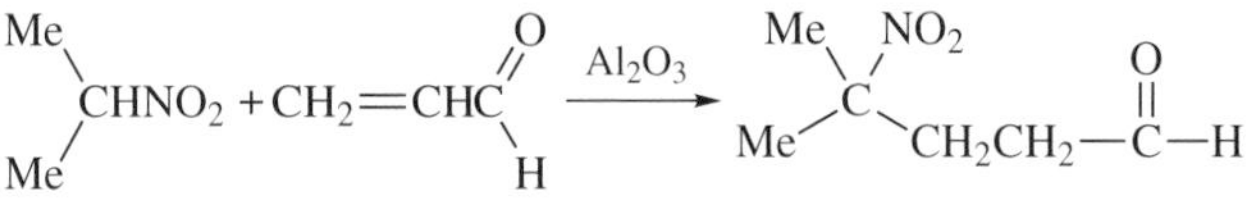

$$\mathrm{Me_2CHNO_2} + \mathrm{CH_2{=}CHCHO} \xrightarrow{Al_2O_3} \mathrm{Me_2C(NO_2)CH_2CH_2{-}C(=O){-}H}$$

(b)

$$\mathrm{Me_2C{=}C(Me)C(=O)Me} + \mathrm{CH_2(CO_2Et)_2} \xrightarrow{\text{NaOEt}} \left[\mathrm{MeC(=O)CH(Me)C(Me)_2CH(CO_2Et)_2}\right] \longrightarrow$$

ethyl 2,2,3-trimethyl-4,6-dioxocyclohexanecarboxylate (ring: $C(Me)_2$, $CH(CO_2Et)$, C=O, CH_2, C=O, $CH(Me)$)

(c)

$$\mathrm{Me_2C{=}O} + \mathrm{BrCH(Me)CO_2Et} \xrightarrow{\text{Zn}} \mathrm{Me_2C(OH){-}CH(Me)CO_2Et}$$

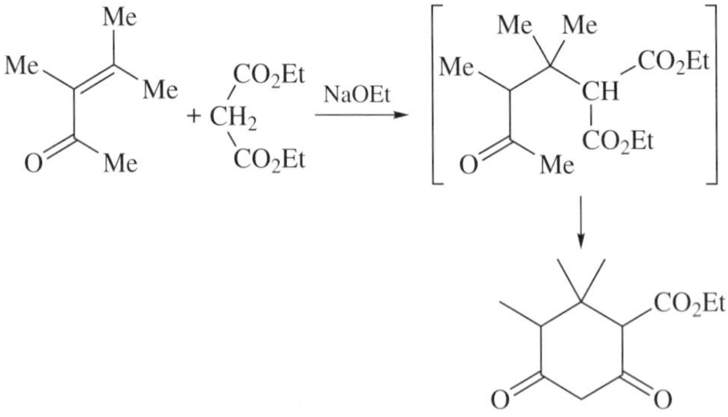

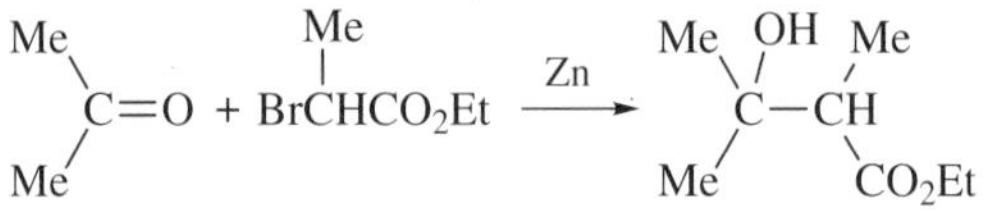

(d) $(EtO_2C)_2CH_2 + CH(CO_2Et){=}CHCO_2Et \xrightarrow{NaOEt} (EtO_2C)_2CHCH(CO_2Et)CH_2CO_2Et$

(e) $MeCOCH_2CO_2Et + MeI \xrightarrow{NaOEt} MeCOCH(Me)CO_2Et \xrightarrow{BrCH_2CH{=}CH_2,\ NaOEt} MeCOC(Me)(CH_2CH{=}CH_2)CO_2Et \xrightarrow{NaOH} MeCOC(Me)(CH_2CH{=}CH_2)CO_2H \xrightarrow{H_3O^+} MeCOCH(Me)CH_2CH{=}CH_2$

(f) $CH_2BrCH_2CH_2Br + CH_2(CO_2Et)_2 \xrightarrow{NaOEt}$ 1,1-cyclobutane($CO_2Et)_2$ $\xrightarrow{1.\ NaOH;\ 2.\ H_3O^+}$ cyclobutane-CO_2H

(g) $CH_2(CO_2Et)_2 \xrightarrow{MeI,\ NaOEt} MeCH(CO_2Et)_2 \xrightarrow{PhCH_2Br,\ NaOEt} MeC(PhCH_2)(CO_2Et)_2 \xrightarrow{1.\ NaOH;\ 2.\ H_3O^+} MeCH(PhCH_2)CO_2H$

(h) $2\,CH_2(CO_2Et)_2 + BrCH_2CH_2Br \xrightarrow{NaOEt} (EtO_2C)_2CHCH_2CH_2CH(CO_2Et)_2 \xrightarrow{1.\ NaOH;\ 2.\ H_3O^+} HO_2CCH_2CH_2CH_2CH_2CO_2H \xrightarrow{H_3O^+,\ EtOH} EtO_2CCH_2CH_2CH_2CH_2CO_2Et \xrightarrow{NaOEt}$ 2-oxocyclopentane-CO_2Et

(i) $MeCOMe + MeCOOEt \xrightarrow{NaOEt} MeCOCH_2COMe \xrightarrow{I_2} MeCOCH(I)COMe$

(j) $C_6H_5CHO + MeCOMe \xrightarrow{NaOH} C_6H_5CH{=}CHCOMe$

3.2.

(a) $MeI \xrightarrow[2.\ ^{14}CO_2]{1.\ Mg} Me^{14}CO_2H \xrightarrow{H_3O^+,\ EtOH} Me^{14}CO_2Et \xrightarrow[NaOEt]{(CO_2Et)_2} EtO_2CC(=O)CH_2{}^{14}CO_2Et \xrightarrow{NaBH_4} EtO_2CCH(OH)CH_2{}^{14}CO_2Et \xrightarrow{NaOH} HO_2CCH(OH)CH_2{}^{14}CO_2H$

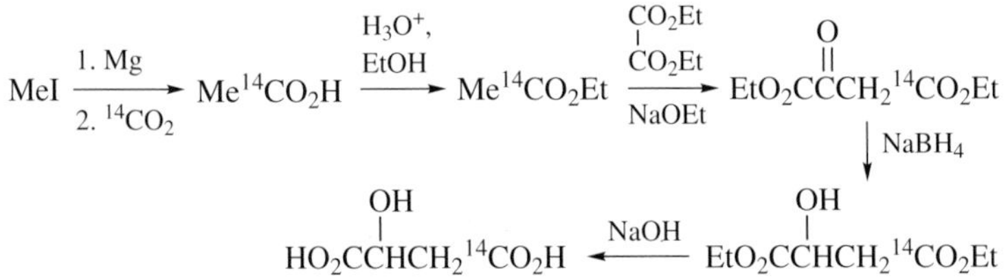

(b) $MeCH_2I \xrightarrow[2.\ ^{14}CO_2]{1.\ Mg} MeCH_2{}^{14}CO_2H \xrightarrow{H_3O^+,\ EtOH} MeCH_2{}^{14}CO_2Et \xrightarrow[NaOEt]{MeCH{=}CHCO_2Et} EtO_2{}^{14}C{-}CH(Me){-}CH(Me){-}CH_2{-}CO_2Et \xrightarrow{NaOH} HO_2{}^{14}C{-}CH(Me){-}CH(Me){-}CH_2{-}CO_2H$

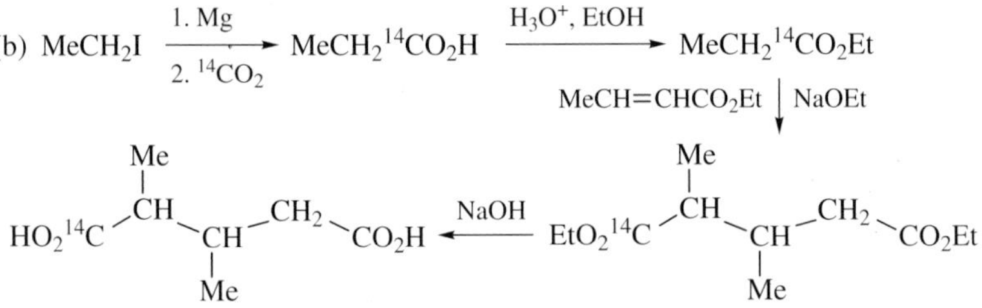

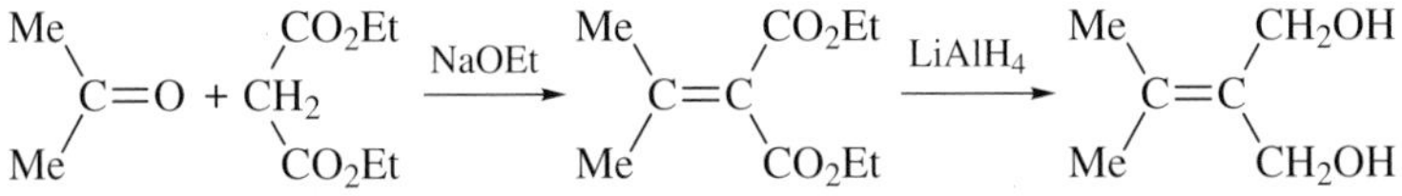

3.3.

(a) $CH_2(CO_2Et)_2 \xrightarrow[2\ MeI]{2\ NaOEt} Me_2C(CO_2Et)_2 \xrightarrow{LiAlH_4} Me_2C(CH_2OH)_2$

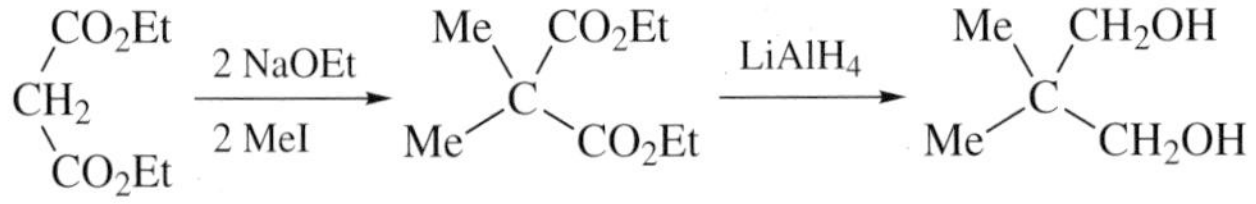

(b) $Me_2C{=}O + CH_2(CO_2Et)_2 \xrightarrow{NaOEt} Me_2C{=}C(CO_2Et)_2 \xrightarrow{LiAlH_4} Me_2C{=}C(CH_2OH)_2$

(c) $MeC(=O)CH_2CO_2Et + 2EtI \xrightarrow{2\ NaOEt} MeC(=O)C(Et)_2CO_2Et \xrightarrow[2.\ H_3O^+]{1.\ NaOH} MeC(=O)CH(Et)_2$

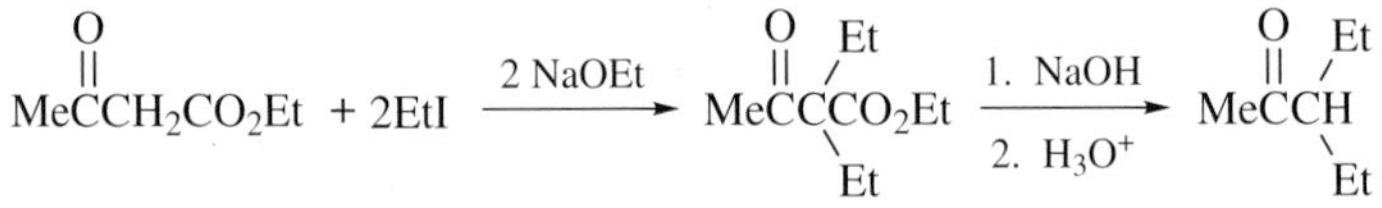

(d) $MeC(=O)CH{=}CH_2 + MeC(=O)CH_2CO_2Et \xrightarrow{NaOEt}$ 3-methyl-4-(ethoxycarbonyl)cyclohex-2-en-1-one (CO_2Et)

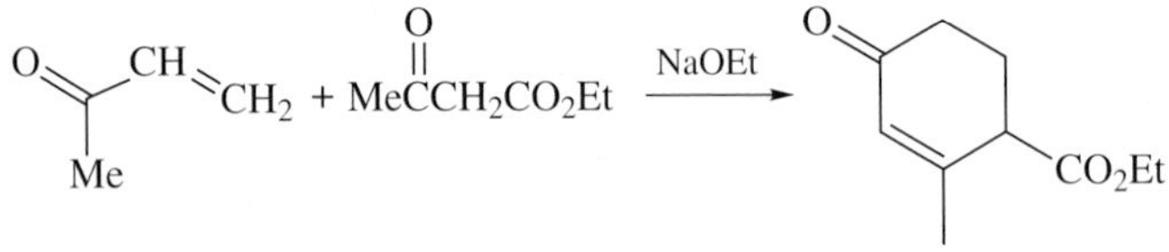

Chapter 4

4.1.

(a) 1,3,5-trihydroxybenzene (OH, HO, OH) $+ PrCN \xrightarrow[HCl]{ZnCl_2}$ 2,4,6-trihydroxyphenyl–C(=O)–Pr (OH, HO, OH)

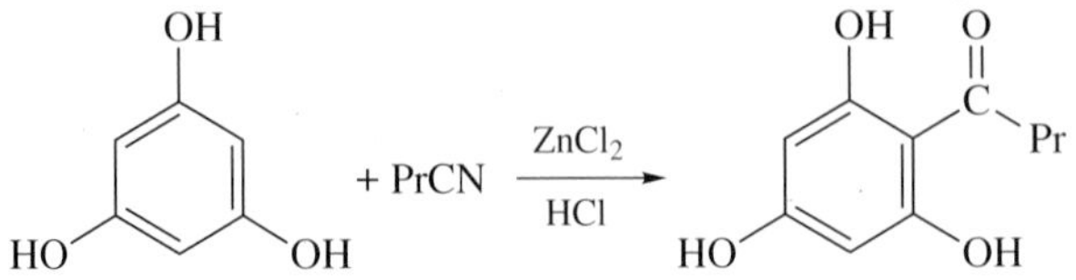

(b) OMe-benzene $\xrightarrow[\text{AcOH}]{HNO_3}$ 4-nitroanisole (OMe, NO_2) $\xrightarrow[\text{ZnCl}_2\text{, HCl}]{\text{HCHO}}$ OMe, CH_2Cl, NO_2

(c) HO-phenyl + $HO-C(Me)_2CH_2CH_2Me$ $\xrightarrow{AlCl_3}$ HO-phenyl-$C(Me)_2CH_2CH_2Me$

(d) Cl-phenyl + maleic anhydride $\xrightarrow{AlCl_3}$ Cl-phenyl-$C(=O)CH{=}CHCO_2H$

(e) Cl-phenyl + CCl_3CHO $\xrightarrow{\text{conc. } H_2SO_4}$ Cl-phenyl-CH(CCl_3)-phenyl-Cl $\xrightarrow{\text{NaOH}}$ Cl-phenyl-C($=CCl_2$)-phenyl-Cl

4.2.

Benzene + Me_2CHCl $\xrightarrow{AlCl_3}$ phenyl-CH(Me)Me $\xrightarrow[AlCl_3]{\text{succinic anhydride}}$ $HO_2CCH_2CH_2C(=O)$-phenyl-CH(Me)Me $\xrightarrow{\text{Zn, HCl}}$ $HO_2C(CH_2)_3$-phenyl-isopropyl $\xrightarrow{AlCl_3}$ isopropyl-tetralone $\xrightarrow{\text{MeMgI}}$ tertiary alcohol (Me, OH) $\xrightarrow{\text{Se}}$ 1-methyl-7-isopropylnaphthalene

4.3.

(a) 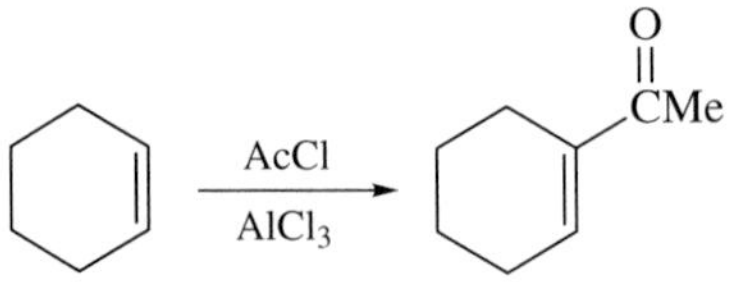 AcCl, $AlCl_3$ → CMe (O)

(b) 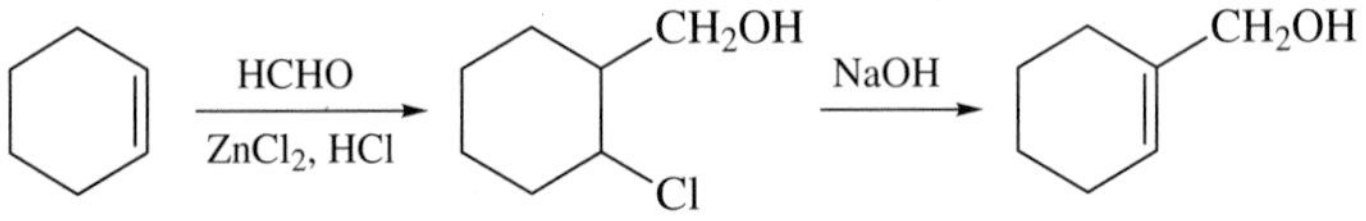 HCHO, $ZnCl_2$, HCl → CH_2OH, Cl; NaOH → CH_2OH

Chapter 5

5.1.

(a) 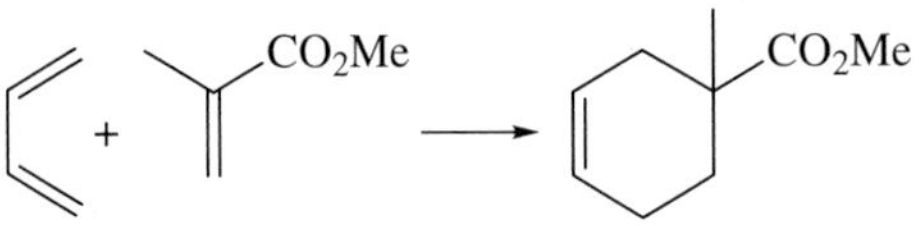 + CO_2Me → CO_2Me

(b) 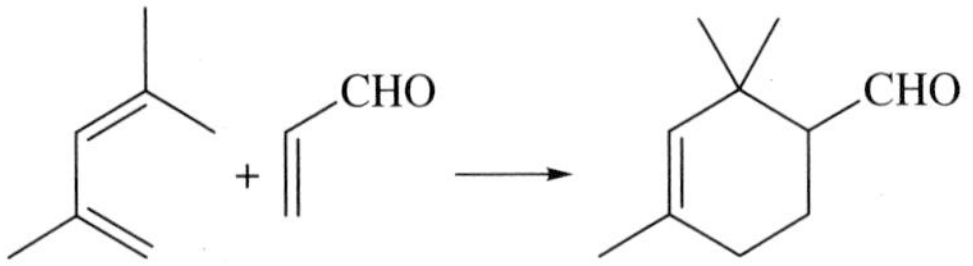 + CHO → CHO

(c) 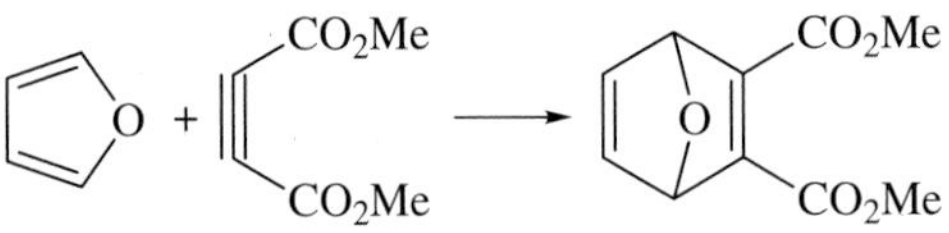 O + CO_2Me, CO_2Me → O, CO_2Me, CO_2Me

(d) 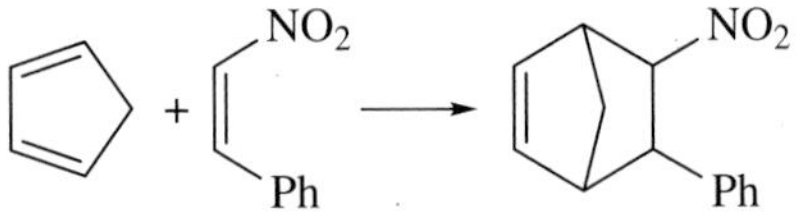 + NO_2, Ph → NO_2, Ph

(e) 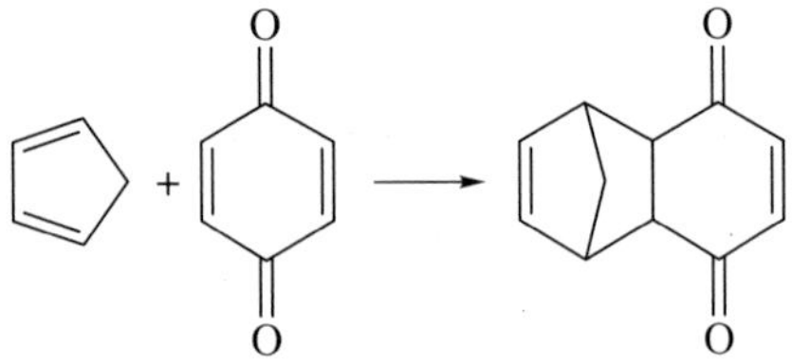 + O, O → O, O

(f) 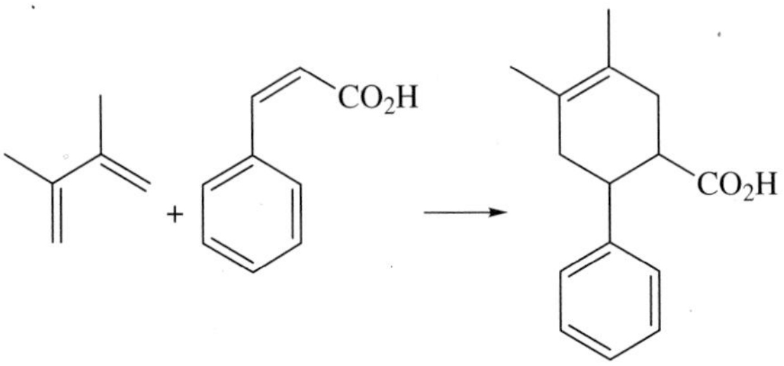 + CO_2H → CO_2H

5.2.

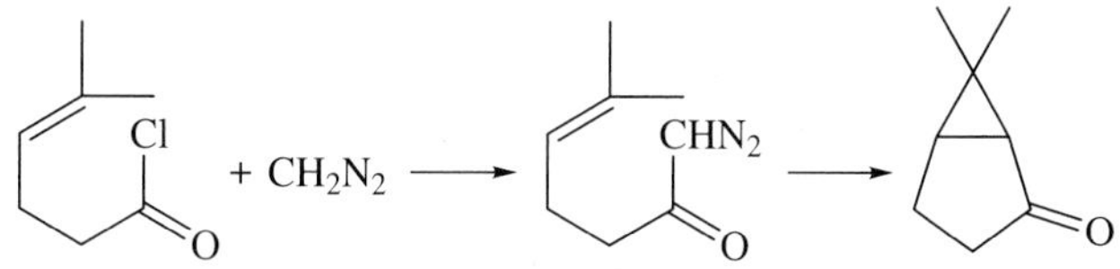

Cl + CH_2N_2 ⟶ CHN_2 ⟶

O O O

5.3.

(a) CO_2Me H

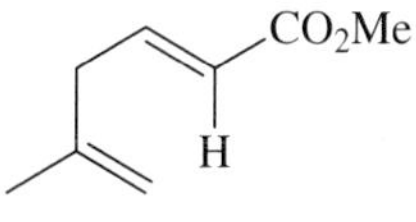

(b) OMe O

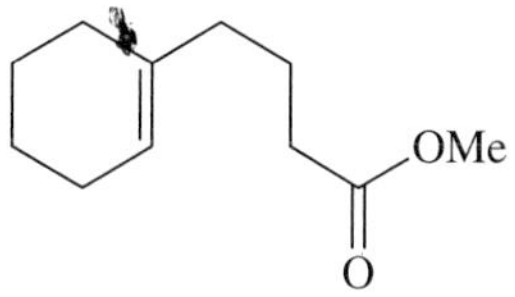

5.4.

(a) CH_2CH_2CN

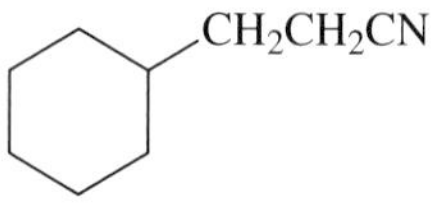

(b) O O $BnOCH_2$ CH_2 CH_2 CH_2 O C Me

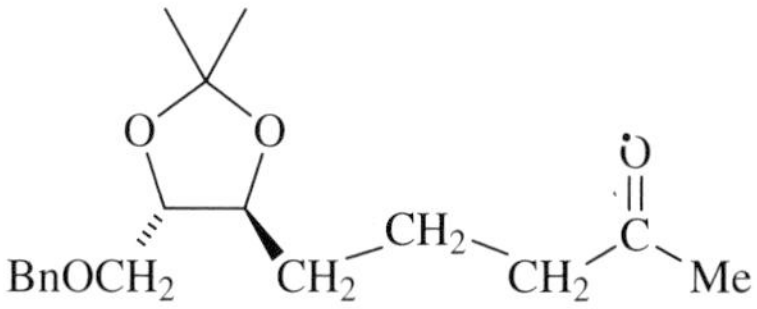

(c) H H O OEt

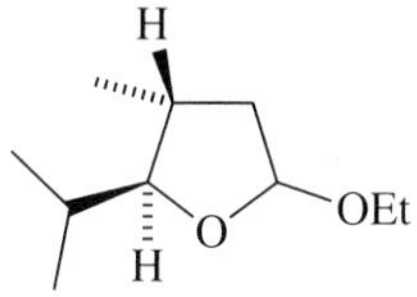

(d) OMe O O O

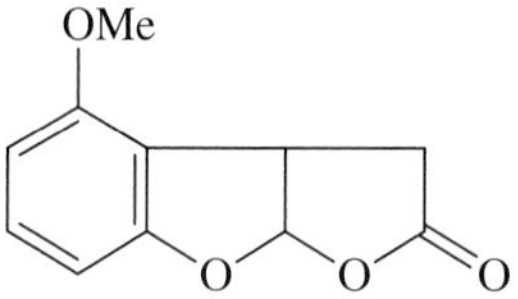

5.5.

O O Me Br Me $Bu_3Sn^{\bullet}$ O O O O

$Bu_3Sn^{\bullet}$

O O Bu_3SnH O O

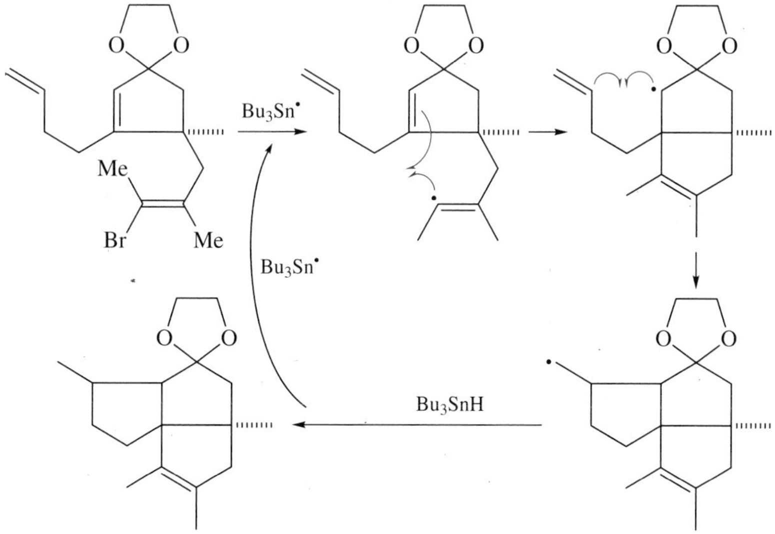

Chapter 6

6.1.

(a)

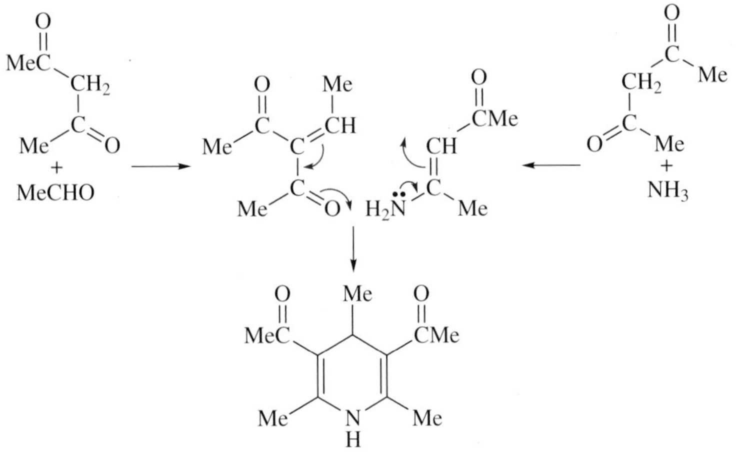

O, MeC, CH_2, Me, C, O
+
MeCHO ⟶

O, Me, Me, C, C, CH, Me, C, O — $H_2\ddot{N}$, C, Me, CH, O, CMe

⟵ O, C, Me, CH_2, O, C, Me
+
NH_3

↓

O, MeC, Me, O, CMe, Me, N, H, Me

(b) $MeCCH_2CO_2Et$ (O) —(1. HNO_2; 2. Zn, AcOH)⟶ $MeCCHCO_2Et$ (O; NH_2)

↓ EtO_2C, CH_2, C, Me, O

EtO_2C, Me, Me, N, H, CO_2Et

(c) NH_2, C, O, NH_2 + O, EtOC, CH_2, C, O, Me ⟶ O, HN, O, N, H, Me

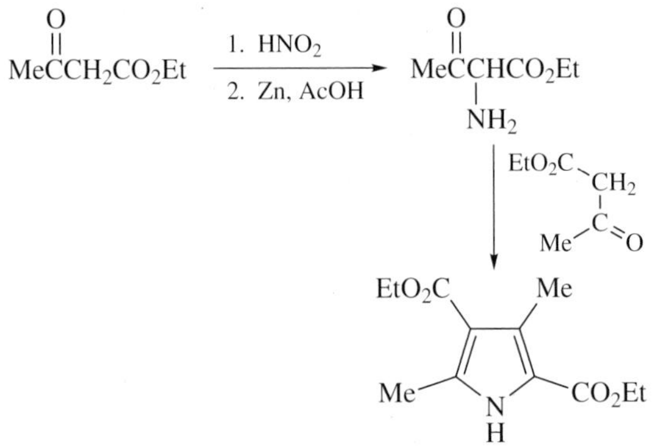

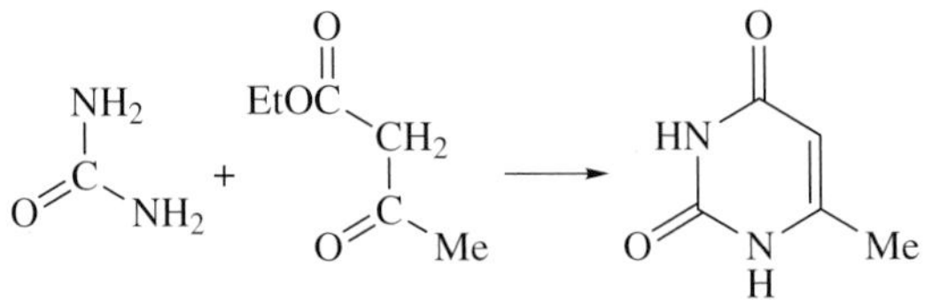

(d) NH_2, NH_2 + O, C, Me, O, C, Me ⟶ N, Me, N, Me

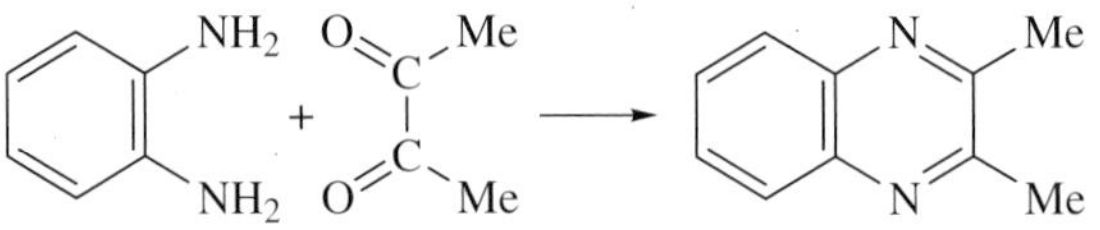

(e) 3-Methoxyphenethylamine $\xrightarrow{AcCl}$ N-acetyl derivative ($O{=}C(Me)NH$) $\xrightarrow{H_3PO_4}$ 6-methoxy-1-methyl-3,4-dihydroisoquinoline

MeO, NH_2, AcCl, MeO, NH, $O{=}C$, Me, H_3PO_4, MeO, N, Me

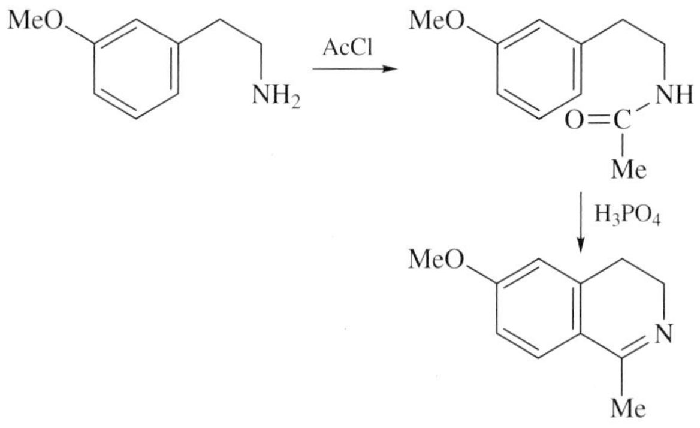

(f) $H{-}C({=}S){-}NH_2$ + $MeC({=}O){-}CHCl{-}CO_2Et$ ⟶ thiazole (4-Me, 5-CO_2Et)

NH_2, H, C, S, +, O, C, Me, CHCl, CO_2Et, Me, N, S, CO_2Et

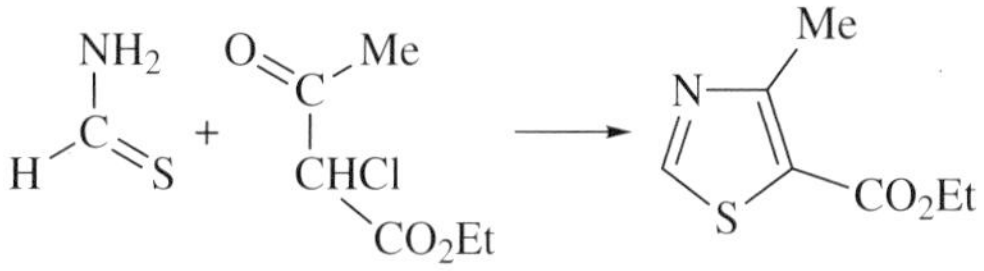

6.2. (i) $NC{-}CH_2{-}CO_2Et$; (ii) HNO_2; (iii) Sn; (iv) HCl; (v) HCO_2H.

6.3.

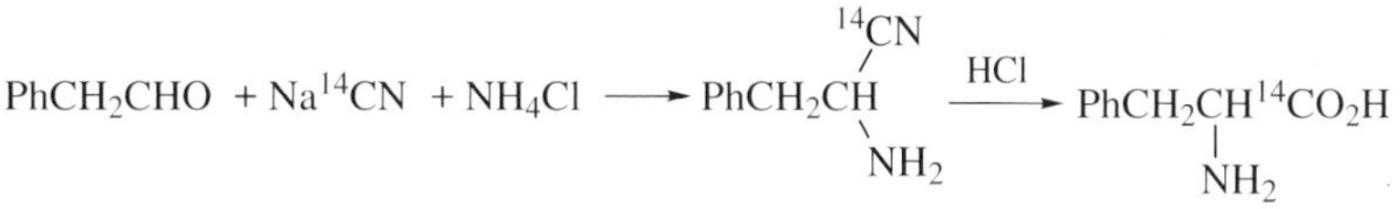

$$PhCH_2CHO + Na^{14}CN + NH_4Cl \longrightarrow PhCH_2CH(^{14}CN)(NH_2) \xrightarrow{HCl} PhCH_2CH(NH_2)^{14}CO_2H$$

6.4.

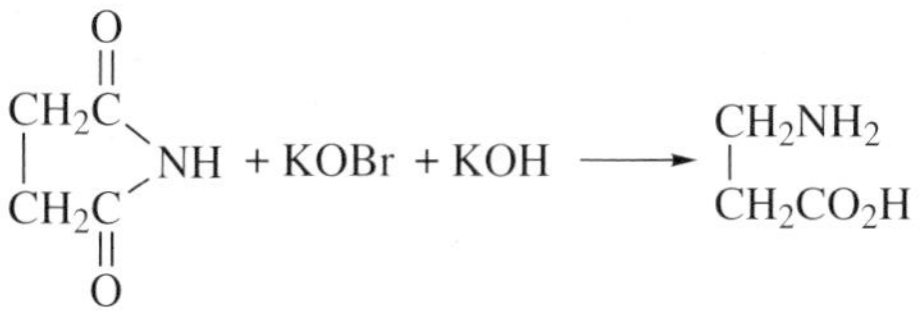

$$\begin{matrix} CH_2C(=O) \\ | \\ CH_2C(=O) \end{matrix}\!\!>NH + KOBr + KOH \longrightarrow \begin{matrix} CH_2NH_2 \\ | \\ CH_2CO_2H \end{matrix}$$

6.5.

$$H_2NCH_2CH_2CH_2Br + CS_2 \longrightarrow$$ 5,6-dihydro-4H-1,3-thiazine-2-thiol (ring with S, N; substituent SH)

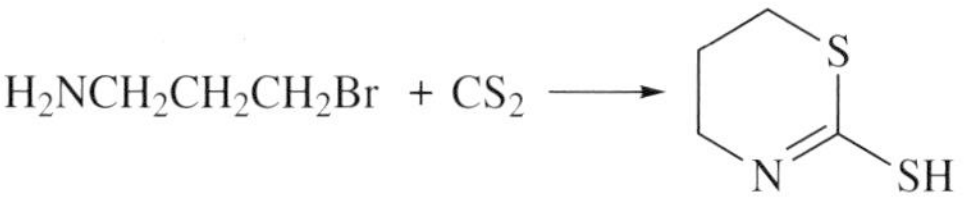

Chapter 7

7.1.

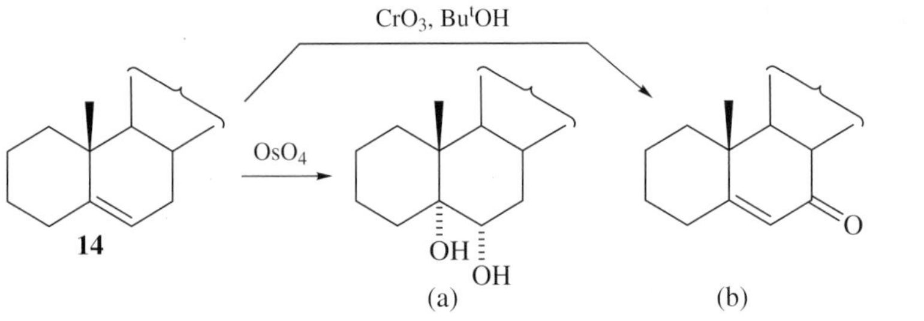

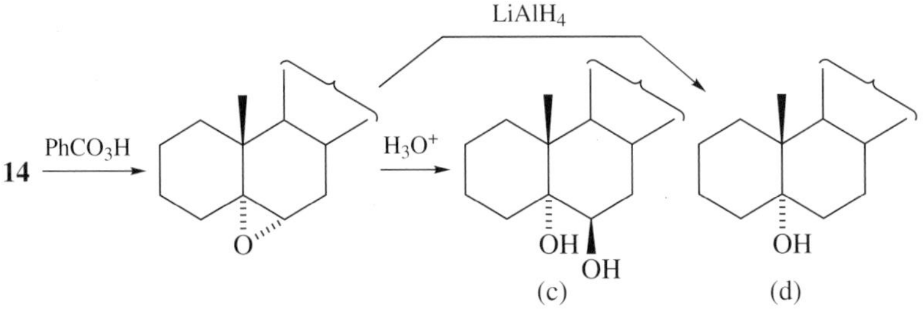

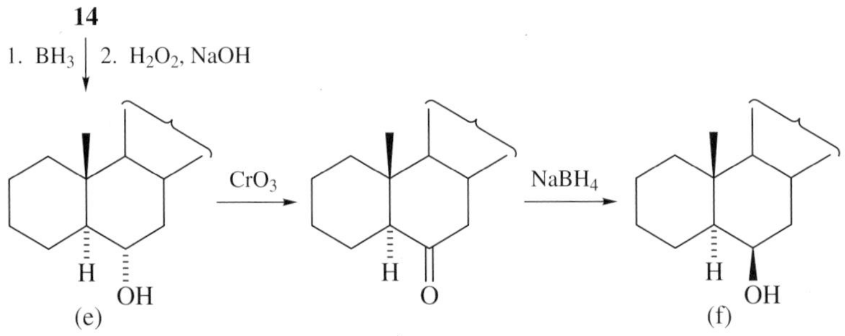

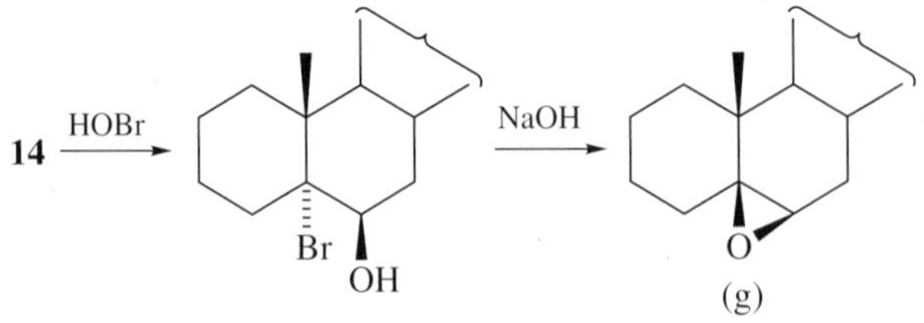

CrO_3, Bu^tOH

14 $\xrightarrow{OsO_4}$ (a) OH OH ; (b) O

$LiAlH_4$

14 $\xrightarrow{PhCO_3H}$ O $\xrightarrow{H_3O^+}$ (c) OH OH ; (d) OH

14

1. BH_3 2. H_2O_2, NaOH

(e) H OH $\xrightarrow{CrO_3}$ H O $\xrightarrow{NaBH_4}$ (f) H OH

14 $\xrightarrow{HOBr}$ Br OH $\xrightarrow{NaOH}$ (g) O

7.2.

(a) HO 15 $\xrightarrow{PhCO_3H}$ HO O

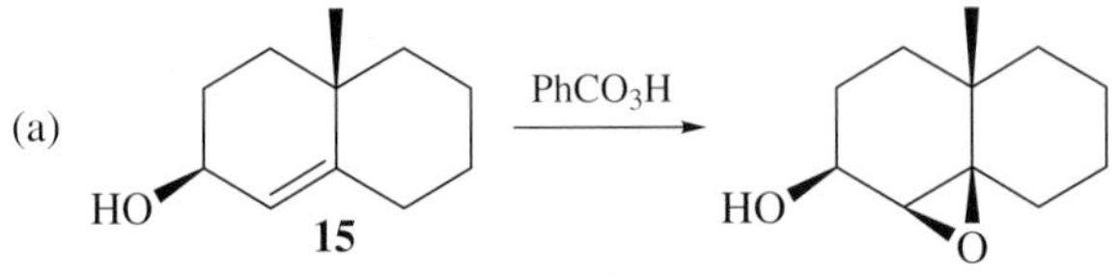

(b) **15** $\xrightarrow{\text{TBDMSCl}}$ TBDMSO– (decalin alkene) $\xrightarrow{\text{PhCO}_3\text{H}}$ TBDMSO– (epoxide) $\xrightarrow{\text{Bu}_4\overset{+}{\text{N}}\text{F}^-}$ HO– (epoxide)

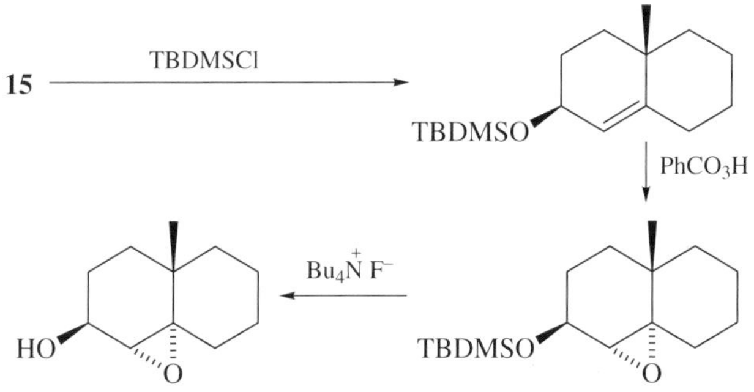

7.3.

(a) $MeC{\equiv}CCH_2OH \xrightarrow{H_2,\ Pd/BaCO_3}$ (H)(Me)C=C(H)(CH_2OH) (cis)

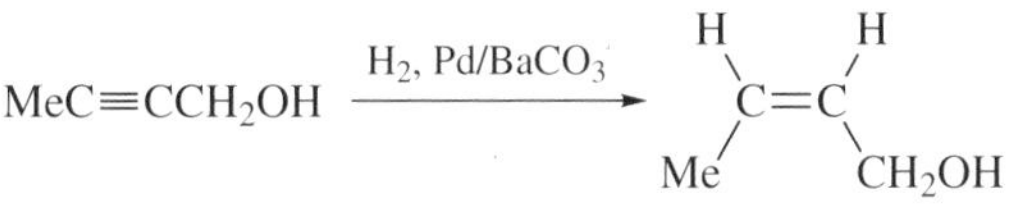

(b) $MeC{\equiv}CCH_2OH \xrightarrow{LiAlH_4 \text{ or } Li,\ NH_3}$ (Me)(H)C=C(H)(CH_2OH) (trans)

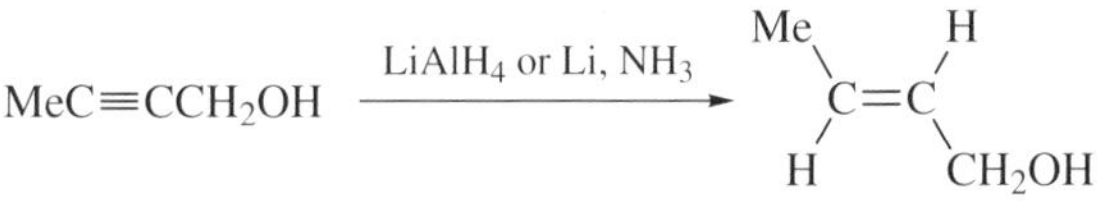

7.4.

(a) $EtO_2CCH_2CH_2{}^{14}CO_2H \xrightarrow{SOCl_2} EtO_2CCH_2CH_2{}^{14}COCl$

1. H_2, $Pd/CaCO_3$ 2. $NaBH_4$

$EtO_2CCH_2CH_2{}^{14}CH_2OH$

(b) $EtO_2CCH_2CH_2{}^{14}CO_2H \xrightarrow{LiAlH_4} HOCH_2CH_2CH_2{}^{14}CO_2H$

H_3O^+, EtOH

$HOCH_2CH_2CH_2{}^{14}CO_2Et$

7.5.

(a) 1-methylcyclohexene (Me) $\xrightarrow{HBr}$ 1-bromo-1-methylcyclohexane (Me, Br)

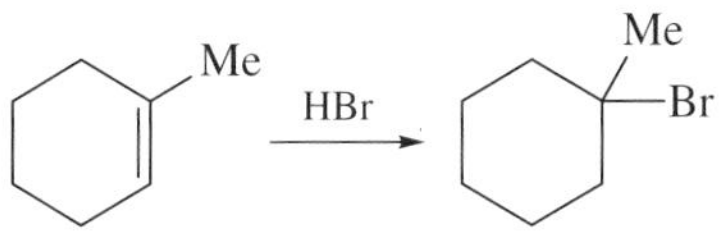

(b) 1-methylcyclohexene (Me) $\xrightarrow{NBS}$ 3-bromo-1-methylcyclohexene (Me, Br)

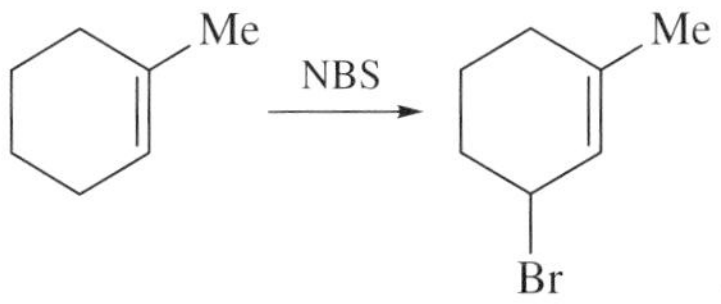

(c) 1-methylcyclohexene (Me) $\xrightarrow[2.\ H_2O_2,\ NaOH]{1.\ BH_3}$ trans-2-methylcyclohexanol (Me, H, OH) $\xrightarrow{PBr_3}$ (Me, H, Br)

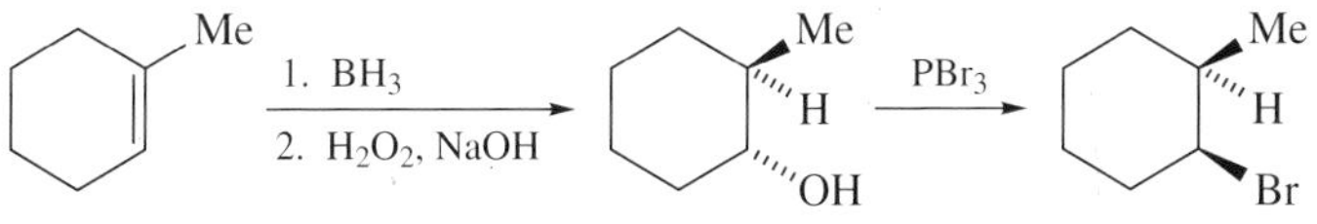

(d) 1-methylcyclohexene $\xrightarrow{\text{Br, AcOH}}$ 1,2-dibromo-1-methylcyclohexane (Me, Br, Br)

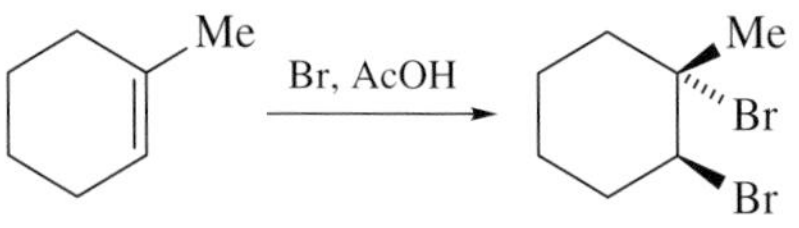

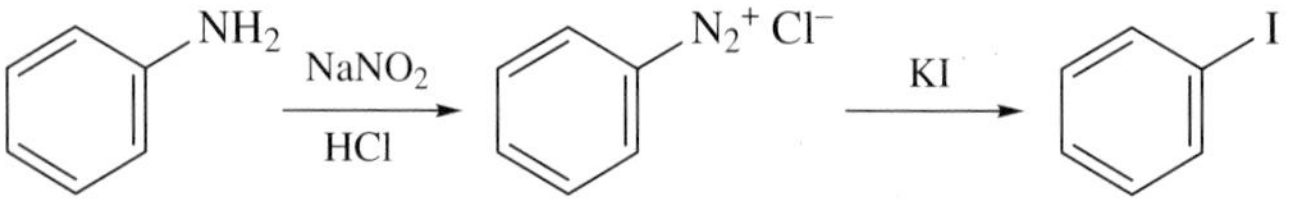

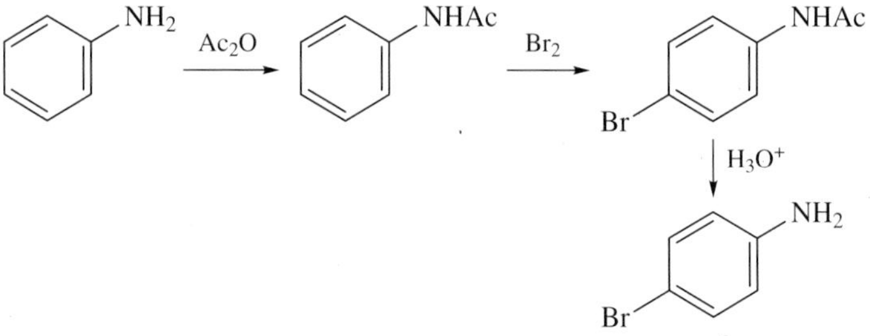

7.6.

(a) aniline (NH_2) $\xrightarrow[\text{HCl}]{NaNO_2}$ $N_2^+\ Cl^-$ $\xrightarrow{\text{KI}}$ iodobenzene (I)

(b) aniline (NH_2) $\xrightarrow{Ac_2O}$ NHAc $\xrightarrow{Br_2}$ Br, NHAc $\xrightarrow{H_3O^+}$ Br, NH_2

(c) phenol (OH) $\xrightarrow{H_2SO_4}$ OH, HO_3S, SO_3H $\xrightarrow{Br_2}$ OH, HO_3S, Br, SO_3H $\xrightarrow{H_3O^+}$ OH, Br

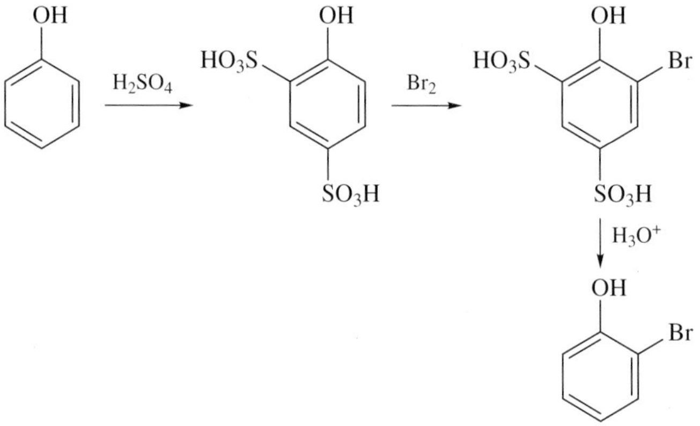

(d) NH_2, CO_2H $\xrightarrow{\text{ICl}}$ NH_2, I, CO_2H $\xrightarrow[\text{EtOH, Cu}]{NaNO_2\text{, HCl}}$ I, CO_2H

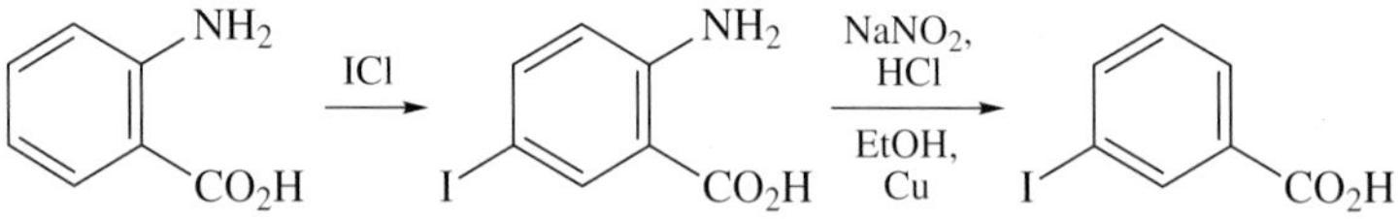

7.7.

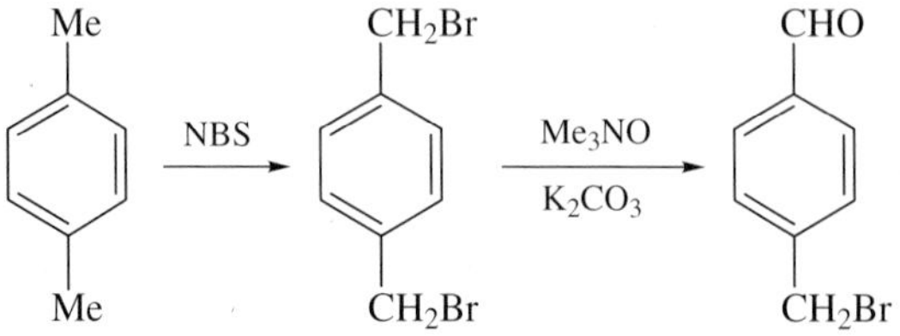

Me, Me $\xrightarrow{\text{NBS}}$ CH_2Br, CH_2Br $\xrightarrow[K_2CO_3]{Me_3NO}$ CHO, CH_2Br

7.8. The first step proceeds with inversion of configuration whilst the second step proceeds *via* an α-lactone with retention of configuration.

Chapter 8

8.1.

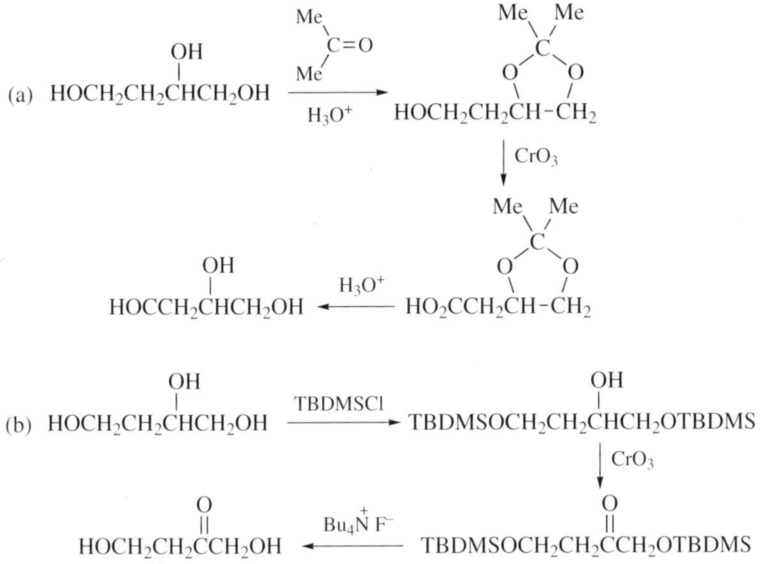

(a) $HOCH_2CH_2CH(OH)CH_2OH \xrightarrow[H_3O^+]{Me_2C=O}$ $HOCH_2CH_2CH-CH_2$ (acetonide: O–C(Me)(Me)–O bridging CH and CH_2) $\xrightarrow{CrO_3}$ $HO_2CCH_2CH-CH_2$ (acetonide) $\xrightarrow{H_3O^+}$ $HOCCH_2CH(OH)CH_2OH$

(b) $HOCH_2CH_2CH(OH)CH_2OH \xrightarrow{TBDMSCl}$ $TBDMSOCH_2CH_2CH(OH)CH_2OTBDMS \xrightarrow{CrO_3}$ $TBDMSOCH_2CH_2C(=O)CH_2OTBDMS \xrightarrow{Bu_4\overset{+}{N}F^-}$ $HOCH_2CH_2C(=O)CH_2OH$

8.2.

OHC-substituted decalone $\xrightarrow[H_3O^+]{MeOH}$ $(MeO)_2CH$-substituted decalone $\xrightarrow{MeMgI}$ $(MeO)_2CH$-substituted decalol (Me, OH) $\xrightarrow{H_3O^+}$ OHC-substituted decalol (Me, OH)

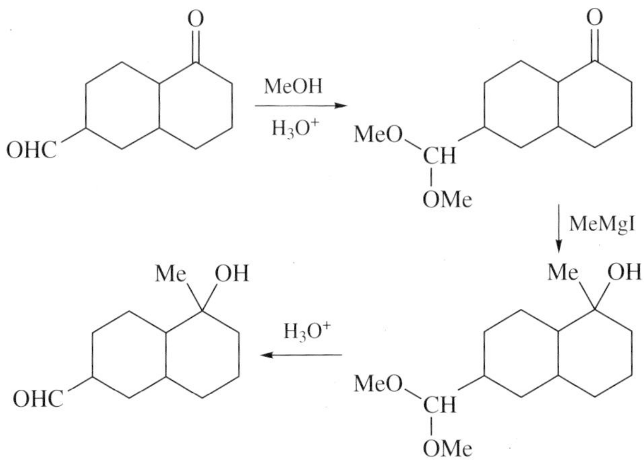

8.3.

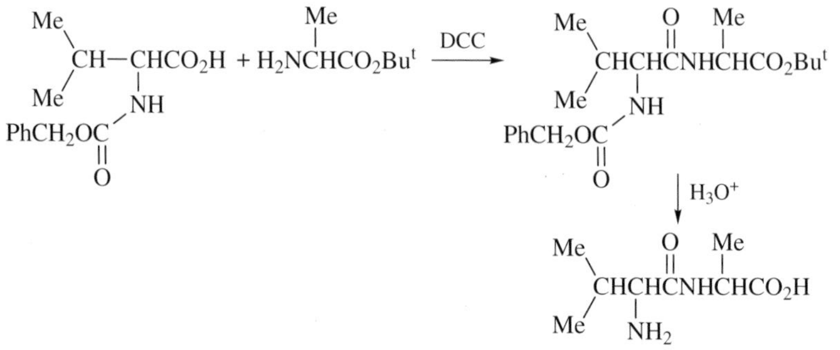

8.4.

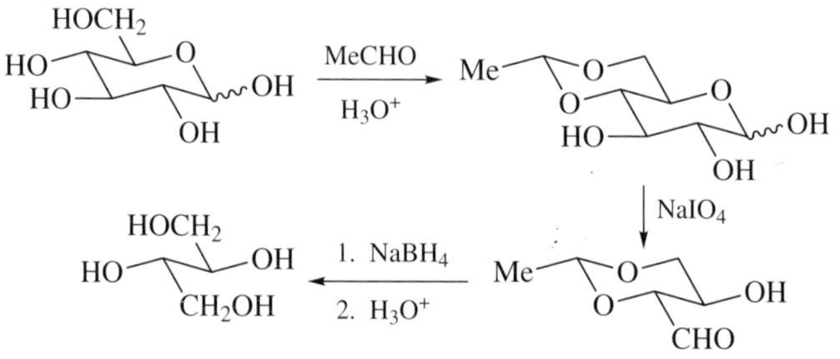

Subject Index

Acetals 54
Acetamidomalonate reaction 86, 87
Acetonides 129
Acetylides 20, 21
Acylium ions 54, 55
Acyloin condensation 66, 67
Aldol condensation 39, 40
Alkene metathesis 70
Allyl ethers 127
Amino acid synthesis 42, 85
Amino group protection 132
Anti-Markownikoff hydration 111
Arndt–Eistert reaction 69
Azlactones 42, 85
Azobisisobutyronitrile 64, 104

Baeyer–Villiger oxidation 102
Baldwin's rules 47
9-BBN 111
Beckmann rearrangement 84
Benzyl ethers 127
Benzyloxycarbonyl (Cbz) 134
BINAP 106
Birch reduction 104, 106
Boranes 27, 110
Bromination 116
N-Bromosuccinimide (NBS) 115, 118

Cadiot–Chodkiewicz coupling 65, 66
Carbanions 32–50
Carbenes 69
Carbocations 54–62
Carbonates 129
Carbonyl group protection 131
Carbopalladation 16
Carboxylic acid protection 130
Chiral auxiliary 49, 50
Chloroanil 97
m-Chloroperbenzoic acid 101, 102
Chlorotrimethylsilane 12
Chromium trioxide 97, 102
Cinnamic acid 41
Claisen rearrangement 73
Clemmensen reduction 107
Combinatorial synthesis 136
Convergent synthesis 4
Cope rearrangement 73
Curtius rearrangement 84
Cyclization 47
Cyclocitral 55
Cyclopentadienyl anion 10

Dakin oxidation 103
DDQ 97
Dehydrogenation 96
Dess–Martin periodinane oxidation 99
Diazomethane 69, 70, 130
Dibenzoyl peroxide 64
Dicyclohexylcarbodiimide (DCC) 135
Dieckmann cyclization 43
Diels–Alder reaction 71
Diene 71
Dienophile 71
Diethylaminosulfur trifluoride (DAST) 120
β-Diketones 34, 88
Dimedone 45
Dimethylsulfonium methiodide 25
Dimethyl sulfoxide 25, 34, 98
Dimethylsulfoxonium methylide 25
Dimethyltitanocene 24
1,3-Dioxolanes 131
Disconnection 2
Dissolving metal reductions 106

Dithioacetals 26, 132
Doebner–Knoevenagel condensation 41
Doebner–von Miller synthesis 70

Enamines 35, 49, 88
Enantiospecific 6, 49, 101
Ene reaction 73
Enol ethers 35, 47, 48
Enolate anions 32–33, 48, 49
Enols 32, 33
Erlenmeyer synthesis 42, 85
Eschenmoser's salt 57
Ethyl acetoacetate 34, 36, 37, 43, 44, 149
Eudesmol 143

9-Fluorenylmethoxycarbonyl (Fmoc) 134
Fischer indole synthesis 90
Fluorination 119
Friedel–Crafts acylation 57, 59, 61, 145, 147
Friedel–Crafts alkylation 55, 59

Gabriel synthesis 81
Grignard reagents 10, 11–14, 58, 104, 143

Halogenation 111–121
Hantzsch synthesis 88, 89
Heck reaction 17
Hofmann degradation of amides 84
Hunsdiecker reaction 108
Hydrazine 101
Hydroboration 110, 144
Hydrogenation 104
Hydrogenolysis 103
1-Hydroxybenzotriazole ester 136
Hydroxy group protection 127
N-Hydroxysuccinimide ester 136
Hypochlorous acid 115

Ibuprofen 57
Iminium salts 57, 58
Iodination 117
Ireland–Claisen rearrangement 73

Jones' reagent 97

β-Keto acid 37
β-Keto esters 34, 36, 88
Knorr synthesis of pyrroles 89, 90
Kolbe electrolysis 65
Kyodai nitration 79

Leuckart reductive amination 82
Lewis acids 54, 57, 79, 128
Lindlar catalyst 105
Lithium aluminium hydride 103, 104, 107
Lithium diisopropylamide 36, 50, 74
Lössen rearrangement 84

Malonate esters 34, 45, 86, 88
Mannich reaction 56
McMurry coupling 66, 106
Meerwein–Ponndorf reduction 110
Merrifield synthesis 136
Methoxymethyl ethers 128
Methyl ethers 127
Methylcyclohexanone 37
Michael addition 40, 45, 46, 48, 89, 145, 146
Mitsunobu reaction 113, 114

Nitration 79, 80
Nitriles 20, 34, 81
Nitro compounds 34, 42
Nitrophenyl ester 136
Nitrosation 80, 82, 86
Nucleophilic substitution 80, 112

Organocadmium compounds 14
Organocuprates 15, 150
Organolithium compounds 12, 15
Organometallic reagents 9–28
Organopalladium compounds 16
Orthometallation 18
Orton rearrangement 116
Osmium tetroxide 107
Oxaphosphetanes 23
Oxidation 96–103
Ozone 101

Palladium 104
Pentafluorophenyl ester 136
Peptide synthesis 133
Perkin condensation 41
Peterson reaction 27
Phenol coupling 65, 66, 100, 146
Phosphonate 24, 34
Pinacol coupling 66, 67
Pinacol–pinacolone rearrangement 58, 59
Platinum 104
Potassium *t*-butoxide 36
Potassium permanganate 99
Prins reaction 56, 119
Prostaglandins 150
Protecting groups 6, 126–139
Pschorr synthesis 65, 66
Pyrimidine 88
Pyrroles 87, 89, 90, 133

Raney nickel 104, 105
Reformatsky reaction 10, 49

Regioselective reaction 6
Regiospecific reaction 6, 23, 35
Retrosynthetic analysis 2, 143, 145, 147
Robinson ring extension 45, 143
Rosenmund reaction 105

Sandmeyer reaction 117
Sharpless reagent 101
Silanes 27
Silicon reagents 26
Silver oxide 98
Skraup synthesis 90
Sodium borohydride 107, 143
Sodium–liquid ammonia 104, 106
Sodium periodate 100
Sonogashira coupling 18
Stereoselective 6
Stereospecific 6
Stille coupling 17
Stobbe reaction 41
Strecker synthesis 85
Sulfonamides 92
Sulfoxides 25, 34
Sulfur ylides 25
Suzuki reaction 18
Swern oxidation 98
Synthon 2

Tetrahydropyranyl ethers 128
Thexylborane 111
Thiamine 147
Thionyl chloride 112
Thorpe–Ziegler reaction 43
TPAP 99
Trichloroethyl esters 131
Trimethylamine *N*-oxide 98
Trimethylsilyl ethers 129
Trimethylsilyl iodide 130
Triphenylmethyl ethers 127
Triphenylphosphine 12, 22, 113
Triphosgene 129

Ullmann coupling 65
Urea 88, 147

Vilsmeier reaction 57

Wadsworth–Emmons reaction 24, 150
Wagner–Meerwein rearrangement 58
Wharton reaction 109
Wilkinson's catalyst 105
Wittig reaction 22–25, 143, 150
Wolff–Kischner reduction 109